SOFT COMPUTING

PRENTICE HALL SERIES ON ENVIRONMENTAL AND INTELLIGENT MANUFACTURING SYSTEMS
M. Jamshidi, Editor

Volume 1: *ROBOTICS AND REMOTE SYSTEMS FOR HAZARDOUS ENVIRONMENTS,* edited by M. Jamshidi and P. J. Eicker.

Volume 2: *FUZZY LOGIC AND CONTROL: Software and Hardware Applications,* edited by M. Jamshidi, N. Vadiee, and T. J. Ross.

Volume 3: *ARTIFICIAL INTELLIGENCE IN OPTIMAL DESIGN AND MANUFACTURING,* edited by Z. Dong.

Volume 4: *SOFT COMPUTING: Fuzzy Logic, Neural Networks, and Distributed Artificial Intelligence,* edited by F. Aminzadeh and M. Jamshidi

SOFT COMPUTING

Fuzzy Logic, Neural Networks, and Distributed Artificial Intelligence

EDITORS

Fred Aminzadeh
Unocal Corporation
Mohammad Jamshidi
University of New Mexico

PTR Prentice Hall, Englewood Cliffs, New Jersey 07632

Library of Congress Cataloging-in-Publication Data

```
Soft computing : fuzzy logic, neural networks, and distributed
  artificial intelligence / editors, Fred Aminzadeh, Mohammad
  Jamshidi.
      p.  cm. -- (Prentice Hall series on environmental and
  intelligent manufacturing systems ; v. 4)
         Included bibliographical references and index.
         ISBN 0-13-146234-2
      1. Application software.  2. Fuzzy systems.  3. Neural networks
  (Computer science)  4. Artificial intelligence.  I Aminzadeh,
  Fred.  II. Jamshidi, Mohammad.  III. Series
  QA76.76.A65S657  1994
  006.3--dc20                                              94-4218
                                                              CIP
```

Editorial/production supervision: *Harriet Tellem*
Cover design: *Bruce Kenselaar*
Manufacturing manager: *Alexis Heydt*
Acquisitions editor: *Michael Hays*
Editorial assistant: *Kim Intindola*

©1994 by P T R Prentice Hall
Prentice-Hall, Inc.
A Paramount Communications Company
Englewood Cliffs, New Jersey 07632

The publisher offers discounts on this book in bulk quantities.
For more information contact:

    Corporate Sales Department
    P T R Prentice Hall
    113 Sylvan Avenue
    Englewood Cliffs, New Jersey 07632

    Phone: 201-592-2863
    FAX: 201-592-2249

All rights reserved. No part of this book may be reproduced, in any form or by any means, without permission in writing from the publisher.

Printed in the United States of America
10 9 8 7 6 5 4 3 2 1

ISBN 0-13-146234-2

Prentice-Hall International (UK) Limited, *London*
Prentice-Hall of Australia Pty. Limited, *Sydney*
Prentice-Hall Canada Inc., *Toronto*
Prentice-Hall Hispanoamericana, S.A., *Mexico*
Prentice-Hall of India Private Limited, *New Delhi*
Prentice-Hall of Japan, Inc., *Tokyo*
Simon & Schuster Asia Pte. Ltd., *Singapore*
Editora Prentice-Hall do Brasil, Ltda., *Rio de Janeiro*

DEDICATION:

To Kathleen, Sara, David, and Diana

...FA

To Jila, Ava, and Nima

...MJ

TABLE OF CONTENTS

Foreword ix

Preface xi

List of Contributors xv

Chapter 1
The Evolution of Expert Systems 1
    L.O. Hall and A. Kandel, University of South Florida

Chapter 2
Applications of Fuzzy Expert Systems in Integrated Oil Exploration 29
    F. Aminzadeh, Unocal Corporation

Chapter 3
Hardware Applications of Fuzzy Logic Control 45
    M. Jamshidi, R. Marchbanks, E. Kristjansson, K. Kumbla,
    R. Kelsey, and D. Barak, University of New Mexico

Chapter 4
Fuzzy Expert Systems 99
    R. A. Aliev, Baku University

Chapter 5
Preprocessing Fuzzy Production Rules 109
    M. Schneider and G. Chew, Florida Tech; A. Kandel,
    University of South Florida; and, G. Langholz, Tel Aviv University

Chapter 6
Computational Neural Architectures for Control Applications 121
    L. Jin, M.M. Gupta, and P.N. Nikiforuk
    University of Saskatchewan

Chapter 7
The Application of Artificial Neural Networks in EditingNoisy Seismic Data .. 153
    X. Zhang and Y. Li, Tsinghua University

TABLE OF CONTENTS (cont'd)

Chapter 8
Foundations of Fuzzy Neural Computations 165
  M. Gupta and H. Ding, Univ. of Saskatchewan

Chapter 9
A Control Algorithm for Knowledge-Based Tecture Image Segmentation 201
  Z. Zhang, Z. Lang, and R. E. Scarberry, Medical University
  of South Carolina; M. Simaan, University of Pittsburgh

Chapter 10
Knowledged-Based On-Line Scheduling for Flexible Manufacturing 217
  I. Hatono and H. Tamura, Osaka University

Chapter 11
Coordination of Distributed Intelligent Systems 229
  M. Kamel and H. Ghenniwa, University of Waterloo

Chapter 12
Collaborative Work Based on Multiagent Architectures:
A Methodological Perspective 261
  B. Moulin, Université Laval, and L. Cloutier, Centre de
  Recherche Pour la Défense

Author Index 297

Subject Index 299

FOREWORD

The editors of this volume, Dr. F. Aminzadeh of UNOCAL Corporation and Professor M. Jamshidi of the University of New Mexico, have assembled an authoritative collection of contributions on a wide variety of subjects ranging from coordination of distributed intelligent systems to foundations of fuzzy neural computation. The unifying theme which runs through these contributions is the concept of what is come to be known as soft computing.

What is soft computing? Soft computing (SC)—unlike traditional, hard computing (HC)—is tolerant of imprecision, uncertainty and partial truth. Thus the guiding principle of soft computing is: Exploit the tolerance for imprecision, uncertainty and partial truth to achieve tractability, robustness and low solution cost. Underlying this principle is an obvious and yet frequently neglected fact, namely, that precision carries a cost. What this implies is that to achieve a low cost—or to be able to solve a problem—it is necessary to aim at a solution which is precise enough but no more than necessary. The same applies to uncertainty and partial truth.

A familiar case in point is the problem of parking a car. We can park a car without making any measurements of distance and angle because the final position of the car is not specified precisely. If it were, say to within a fraction of a millimeter and a few seconds of arc, it would take hours of maneuvering and instrumented measurements to solve the problem. Furthermore, the cost of solution would grow geometrically with increase in precision.

Exploitation of the tolerance for imprecision, uncertainty and partial truth plays a key role in data compression, and especially in HDTV, audio recording, speech recognition, image understanding and related fields. But what should be recognized is that the guiding principle of soft computing has much wider implications—implications which cut across disciplinary lines and transcend specific application areas. Actually, SC-based concepts and techniques are already playing an essential role in the conception, design and manufacturing of high MlQ (Machine Intelligence Quotient) products and systems. What this suggests is that eventually the guiding principle of soft computing will be accepted as self-evident truth.

At this juncture, the principal constituents of soft computing (SC) are fuzzy logic (FL), neural network theory (NN) and probabilistic reasoning (PR), with the latter subsuming belief networks, genetic algorithms, chaos theory and parts of learning theory. In the triumvirate of SC, FL is concerned in the main with imprecision, NN with learning and PR with uncertainty. In large measure, FL, NN and PR are complementary rather than competitive. Indeed, what is becoming increasingly clear is that in many cases it is advantageous to employ FL, NN and PR in combination rather than exclusively. In particular, the combination of FL and NN leads to so-called neurofuzzy systems which are rapidly gaining in visibility and are discussed informatively in the chapter authored by M. Gupta and H. Ding.

From a traditional point of view, most of the contributions in this volume fall within the domain of artificial intelligence (AI). There are many observers, myself included, who feel that in the past AI has tied itself too closely to symbol manipulation and conventional, hard computing. Today, we are beginning to see a shift in the orientation of mainstream AI. More specifically, what we see is a significant growth in the number of papers presented at AI conferences or published in AI journals which deal with neural networks, genetic algorithms and belief networks. What this indicates is that AI is nudging closer to soft computing. In fact, it may be argued that it is soft computing—rather than hard computing—that should serve as the foundation for AI. This, perhaps indirectly, is the collective message put forth by the contents of *Soft Computing: Fuzzy Logic, Neural Networks and Distributed Artificial Intelligence*. In so doing, the editors and the authors have made an important contribution to the advancement of soft computing by illuminating the ways in which SC-related techniques can be applied to the solution of real-world problems. For this, they deserve our thanks and congratulations.

Lotfi A. Zadeh

Berkeley, California

PREFACE

Artificial Intelligence (AI) found its way into industrial applications in the early sixties. During the past three decades much effort has been devoted to practical applications of AI. Advances in computer technology have created many futuristic ideas and "science fiction" based dreams to be realized through various AI applications. Combined with the advances in computer hardware, many new mathematical methods and concepts have helped the quantum leap in practical applications of intelligent systems and controllers. As a result, "**Soft Computing**" as a concept was born. Fuzzy logic, neural networks and distributed artificial intelligence are three major components of soft computing.

Soft computing and AI applications have a vast domain and many workers in one discipline may not be aware of the opportunities and challenges in another discipline. This book provides the means for such cross-discipline fertilization of ideas, approaches, and possibilities. Given the rapidly advancing boundaries of fuzzy logic, neural networks, and distributed AI applications, such a book may provide a way to fill gap in familiarizing practitioners in the areas outside their own.

The book is comprised of five segments related to soft computing. It starts with an evolution of the mathematics behind the intelligent systems. It is followed by fuzzy logic and neural network applications. The book concludes with several AI (including distributed AI) applications.

Chapter 1 by Hall and Kandel provides a comprehensive overview of the evolution of expert systems and the new mathematical concepts such as fuzzy logic, neural networks, and genetic algorithms which have revolutionized this field. This chapter offers a simple summary of the fundamentals of terms and concepts providing an excellent background necessary for those who are uninitiated to follow the other chapters. It shows how the use of fuzzy logic and other soft computing concepts have made many expert systems practical by reducing the number of rules (thus computation time) required by the conventional expert systems. This chapter is followed by the application of fuzzy expert systems in integrated oil exploration by Aminzadeh. In this chapter, the fact that use of fuzzy expert systems may be a necessity for the practical use of expert systems in oil exploration is demonstrated through several examples. The importance of the use of fuzzy logic here, in part, is because of the subjective nature of rules governing the oil exploration process. Furthermore, integration of data and rules from different disciplines with various degrees of significance for a specific application is simplified through the use of fuzzy expert systems.

Jamshidi *et al.* describe a number of hardware implementations of fuzzy control systems that are crucial for many fuzzy-based AI systems in Chapter 3. After a brief overview of fuzzy logic, a real-time fuzzy controller and tracker for laser-beam systems is introduced. Additionally, fuzzy controllers for power system generation units, robotic manipulators, controllers for model trains on a circular path, and non-chlorfluoro-carbon air-conditioning and refrigeration systems are discussed.

Chapter 4 by Aliev demonstrates yet another application of fuzzy logic, namely its use in the oil refinery production scheduling problem. Here, it is demonstrated how an intelligent controller based on fuzzy logic simulates an "expert human operator" is handling inexact and subjective "rules" governing the refinery process and loading in different units. Various types of fuzzy production rules and how those rules could be linked to each other in a systematic fashion is discussed in Chapter 5 by Schneider *et al*. They show how this simple procedure renders an architecture that is in a suitable form for many real-time processes, for example those discussed in Chapters 3 and 4.

The next three chapters of the book deal with yet another new addition to AI and intelligent controllers, namely neural networks. This exciting and much researched topic first introduced in the mid fifties and then resurfacing in the late eighties is sure to play a key role in the development of new AI applications in the coming years. Chapter 6 gives an overview of artificial neural networks and a comprehensive introduction to the computational neural architectures for learning control of unknown discrete-time non-linear systems. Mathematical expressions and architectures of both static multilayered feed forward and dynamic recurrent neural networks are described. This chapter is followed by the application of the same artificial neural networks in editing noisy seismic data by Zhang and Li. It is shown how these networks could be used for automatic editing of noisy data, the task that is otherwise performed by a human interpreter. The system goes through a learning process to distinguish between signal and noise. This knowledge is then used in an unsupervised manner to perform the editing task automatically.

Chapter 8 shows how the concepts introduced in Chapters 2 through 5 on fuzzy logic and Chapters 6 and 7 on neural networks could be combined. This integration lays a foundation for a new class of neural networks known as "fuzzy neural networks" (FNN). This chapter builds on the previous work on FNN by presenting an intrinsic interpretation of neural computational mechanisms. It is upon these mechanisms that fuzzy-neural computations are founded. Then a generalized morphology of computational neuron and fuzzy neurons are presented. Three different types of computational fuzzy neural networks are introduced and their intrinsic properties are discussed.

The next four chapters show several examples of knowledge-based systems with or without the use of soft computing. Chapter 9 by Zhang *et al*. describes a control algorithm for knowledge-based 2D and 3D image segmentation systems (ISS). They show how a knowledge-based image (ISS), unlike the conventional ISS, can utilize the information outside a region to classify an image region. This is accomplished by an effective control mechanism that coordinates and balances the segmentation process over the entire image (based on the texture), while providing the means to send information and knowledge from different sources to decision makers at different levels. Hatono and Tamura show a knowledge-based system for on-line scheduling of flexible manufacturing. After an introduction to production system scheduling and its associated problems, a knowledge-based scheduling system for flexible manufacturing under uncertainty is proposed. This is based on a stochastic Petri net that creates, debugs, and evaluates the rules required for scheduling. It is shown that the Petri-net based system improves the operation in a dynamic and uncertain environment such as machine toll failure and processing time variations.

Another relatively new field, distributed artificial intelligence (DAI), is visited in Chapters 11 and 12. Kamel and Ghenniwa address several key issues in DAI. Of particular importance is the issue of coordination among the various AI subsystems. This is accomplished by the introduction of the "Theory of Coordination" to provide a quantitative representation of its main components and the necessary performance measure.

Finally, Chapter 12 by Moulin and Cloutier discusses another aspect of decentralized artificial intelligence, namely, one based on "multi-agent" architecture. This concept is borrowed from DAI, and it is shown how it can be used to support and simulate collaborative work. The Multi-Agent Scenario-Based method of a design involving human and AI agents is described in this chapter. Specifically, it shows how humans and computers (AI systems) can interact in a manner that preserves the autonomy of computers, enabling them to accomplish various tasks on their own and manage their own local data base. This, of course, does not preclude the possibility of having an agent (human or AI subsystem) to require the help of another agent thus initiating a "group activity."

The editors wish to thank William and Virginia Henning of Desert Dreams Publishers of New Mexico for their superb typesetting of the entire manuscript. We wish to also thank Michael Hays and his staff at Prentice Hall for his constant support of this volume and the series. Last, but not least, we wish to thank our families for their continuous understanding and sacrifice during the writing and compiling of this volume.

The reader should expect an exciting and challenging array of possibilities lying with "soft computing." They are exemplified through several informative and excellent papers covering fuzzy systems, neural networks, knowledge-based systems, and distributed artificial intelligence. This is as close as you can get to George Orwell's futuristic book "1984," except this one appears in 1994!

Fred Aminzadeh
Mo Jamshidi

LIST OF CONTRIBUTORS

R. A. Aliev
Dept. of Computer-Aided Control
  Systems
The Azerbaijan State Oil Academy
Baku, AZERBAIJAN

F. Aminzadeh
Unocal Corporation
La Palma Center
5460 East La Palma Avenue
Anaheim, CA 92801, USA

D. Barak
CAD Laboratory for Intelligent and
  Robotic Systems
Dept. of Electrical and Computer
  Engineering
University of New Mexico
Albuquerque, NM 87131-1356, USA

G. Chew
Dept. of Computer Science
Florida Tech.
150 West University Blvd.
Melburne, FL 32901-6988

Louis Cloutier
Division du Commandement et Controle
  Centre de recherche pour la defense,
Valcartier, Courcelette, QC
CANADA, G0A 1RO

Hong Ding
Huazhong University of Science and
  Technology
P.R. CHINA

H. Ghenniwa
Design Engineering
University of Waterloo
Waterloo, Ontario
CANADA N2L 3G1

Madan M. Gupta
Intelligent Systems Research Laboratory
College of Engineering
University of Saskatchewan
Saskatoon, Saskatchewan
CANADA S7N 0W0
Phone: (306) 966-5451
FAX: (306) 966-8710
Email: GUPTAM@sask.usask.ca

L. O. Hall
Dept. of Computer Science & Engineering
University of South Florida
4202 East Fowler Avenue
Tampa, FL 33620-5399, USA

I. Hatono
Dept. of Systems Engineering
Faculty of Engineering Science
Osaka University
1-1 Machikaneyama-cho, Toyonaka
Osaka 560, JAPAN

Mohammad Jamshidi
CAD Laboratory for Intelligent and
  Robotic Systems
Dept. of Electrical & Computer Engineering
University of New Mexico
Albuquerque, NM 87131-1356, USA
Phone: (505) 277-5538
E-mail:jamshid@houdini.unm.edu

Liang Jin
Intelligent Systems Research Laboratory
College of Engineering
University of Saskatchewan
Saskatoon, Saskatchewan
CANADA S7N 0W0
Phone: (306) 966-5451
FAX: (306) 966-8710

M. Kamel
Associate Professor of Systems
Design Engineering
University of Waterloo
Waterloo, Ontario
CANADA N2L 3G1
Phone: (519) 885-1211
FAX: (519) 746-4791

A. Kandel
Dept. of Computer Science & Engineering
University of South Florida
4202 East Fowler Avenue
Tampa, FL 33620-5399, USA

R. Kelsey
Dept. of Computer Science
New Mexico State University
Las Cruces, NM, USA

E. Kristjánsson
Polaroid Corporation
565 Technology Square, #2
Cambridge, MA, 02139 USA

K. Kumbla
CAD Laboratory for Intelligent and
    Robotic Systems
Dept. of Electrical & Computer Engineering
University of New Mexico
Albuquerque, NM 87131-1356, USA

Z. Lang
Dept. of Biostatistics Epidemiology
and Systems Science
Medical University of South Carolina

G. Langholz
Dept. of Electrical Engineering
Tel Aviv University
Tel Aviv, ISRAEL

Y. Li
Dept. of Automation
Tsinghua University
Beijing 100084, CHINA

R. Marchbanks
Los Alamos National Laboratory
Los Alamos, NM, USA

Bernard Moulin
Ingénieur de I'École Centrale de Lyon
    Université Laval
Département d'Informatique
Ste-Foy, Québec, CANADA G1K 7P4
Phone: (418) 656-5580 ou 656-7979
FAX: (418) 656-2324
Email: moulin@vm.ulaval.ca

Peter N. Nikiforuk
Dean of Engineering
College of Engineering
University of Saskatchewan
Saskatoon, Saskatchewan
CANADA S7N 0W0
Phone: (306) 966-5273
FAX: (306) 966-5205

R. E. Scarberry
Dept. of Biostatistics Epidemiology
    and Systems Science
Medical University of South Carolina

M. Schneider
Dept. of Computer Science
Florida Tech.
150 West University Blvd.
Melburne, FL 32901-6988

Marwan Simaan
Dept. of Electrical Engineering
School of Engineering
University of Pittsburgh
348 Benedum Hall
Pittsburgh, PA 15261, USA
Phone: (412) 624-8000
FAX: (412) 624-1108
Email: EEDEPT@EE.PITT.EDU

H. Tamura
Dept. of Systems Engineering
Faculty of Engineering Science
Osaka University
1-1 Machikaneyama-cho, Toyonaka
Osaka 560, JAPAN

Xuegong Zhang
Dept. of Automation
Tsinghua University
Beijing 100084, CHINA

Zhen Zhang
Dept. of Biostatistics Epidemiology
and Systems Science
Medical University of South Carolina

1

THE EVOLUTION OF EXPERT SYSTEMS

L. O. Hall and A. Kandel
University of South Florida
Tampa, Florida, USA

In recent years Expert Systems have evolved mainly into fuzzy expert systems and hybrid systems. Through that evolution, human reasoning and decision processes within an uncertain environment are emulated and manipulated through the tools and techniques provided by fuzzy logic and approximate reasoning. In fuzzy expert systems, consequences are deduced via fuzzy inference mechanisms from a collection of fuzzy IF-THEN rules. The evolution is expanded further by linking the fuzzy expert system with other intelligent agents, such as neural networks, genetic algorithms, and case-base reasoning tools, to provide the user with learning capabilities and increased intelligent power in order to produce systems with high System Intelligence Quotient (SIQ). In this article we describe the classic expert system model, and the stages of the evolutionary process.

1.1 INTRODUCTION

Expert systems are computer programs that emulate the reasoning process of a human expert or perform in an expert manner in a domain for which no human expert exists.

Typically, they reason with uncertain and imprecise information. There are many sources of imprecision and uncertainty. Knowledge they embody is often not exact, in the same way that a human's knowledge is imperfect. The facts or user supplied information is also uncertain.

An expert system is typically made up of at least three parts; an inference engine, a knowledge base, and a global or working memory. The knowledge base contains the expert domain knowledge for use in problem solving. The working memory is used as a scratch pad and to store the information gained from the user of the system. The inference engine uses the domain knowledge together with acquired information about a problem to provide an expert solution.

Almost always a user interface is provided in modern expert systems. This interface is often graphical. It usually allows knowledge and the inference process to be displayed in a number of different ways to help with knowledge acquisition and maintenance. There is often at least a rudimentary explanation capability built into the system. This will lead the user of an expert system through the steps it took to come to a decision. It can be as simple as a trace of rules fired for a rule-based expert system or an analogous case in a case-based expert system.[103]

They typically use some form of high level rules. Blind search of the solution space is avoided and high performance, approaching or surpassing an expert's, is obtained. Reasoning can be done by symbol manipulation. They show some intelligence. Expert systems embody fundamental domain principles and weak reasoning methods. They have difficulty or complexity associated with them. They can reformulate a problem, and some reason about themselves. They can be described as computer programs that use domain knowledge and reasoning techniques to solve problems normally requiring a human expert for their solution. Expert systems may perform a task that humans do not normally perform, such as missile guidance or planning for a robot. An expert system may be able to perform expertly in an area in which there are no human experts.

Expert systems have modeled uncertainty and imprecision in various ways. MYCIN[113] uses certainty factors, while CASNET[122] uses the most significant results of tests. There are expert systems based on probabilistic inference networks, which have a rigorous underlying theory to support uncertain and imprecise information.[95] Interestingly many successful methods of dealing with uncertainty and imprecision in expert systems have been ad hoc, in the sense that there is no underlying theory to support them. They have been validated only via empirical testing.

Expert systems may have several cooperating knowledge sources, which themselves can act as mini-expert systems for a specialized part of a problem. A knowledge source is made up of at least the three basic parts mentioned above. Knowledge sources communicate with each other and with the overall system controller via a device called the blackboard.[28] Messages are posted and received via this blackboard. The information on a blackboard may cause processes to be activated or be used in some other manner. A possible configuration of an expert system is shown in Figure 1.1.

A blackboard may be looked at as a communications coupling device. It provides for loose coupling between several knowledge sources in an expert system. Each knowledge source may be viewed as an individual expert. Therefore an expert system of several knowledge sources is made up of a set of cooperating experts. The experts communicate by writing messages on the blackboard and reading messages from it. A blackboard has a loose resemblance to a mailbox in classic message passing systems.

One important feature that many expert systems incorporate is that of the explanation facility. This facility enables the expert system to explain its reasoning to the user. The user of the system will be able to trace the knowledge used by the system and in some cases determine the motivation for a question that the system has asked. An explanation facility enables a user to determine why information is being asked for and how both intermediate and final conclusions are obtained. It is a very important facility to have when the knowledge base is being debugged. It enables inconsistencies, errors and omissions to be easily identified for correction. A consultation expert system that does not have an explanation facility has little chance of being accepted by experts. They will need to know how a conclusion is arrived at, as well as the reasoning behind the intermediate steps. An expert system which gives advice to a human must be able to satisfy any misconceptions or skepticisms. There are expert systems which do not need an explanation facility, such as Macsyma which solves calculus and algebra problems. However, most interactive expert systems will need some explanation ability to gain user acceptance.

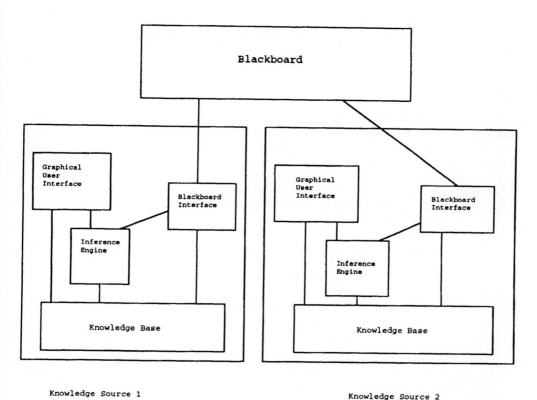

Figure 1.1. A two knowledge source expert system.

In order to understand an expert system's functioning one must see how knowledge is represented in the knowledge base. It is also necessary to understand how the inference engine makes use of the knowledge to come to expert conclusions. Knowledge representation and the inference process will be the subjects of the following sections.

1.1.1 Knowledge Representation

There are many methods of knowledge representation and a thorough description of them is given in.[10, 126, 103] We intend to discuss those that have had a major impact upon expert system development. The three most popular knowledge representation schemes are rules, semantic networks, and frames. Currently, case-based expert systems are of much interest, which store knowledge primarily in the form of past cases.[75, 103]

Under the heading of rule-based systems comes MYCIN which uses production rules as its knowledge representation scheme. The general form of a rule is shown in Figure 1.2.

IF ANTECEDENT **THEN** CONSEQUENT

Figure 1.2. General form of a rule.

Both the antecedent and the consequent are normally given some truth value, often called a certainty. The antecedent is usually some restricted sequence of clauses connected by the connectives and or or. The connectives often serve as min and max operators respectively. Often, rules are set up in some Lisp functional format or a Prolog Horn clause or the antecedent may look like a fully parenthesized logical expression. The consequent may be an action to be taken or a clause, to be added to working memory, which is in some other antecedent. An actual rule is shown in Figure 1.3.

IF class is gymnosperm and leaf shape is needlelike
THEN family is cypress

Figure 1.3. Example rule.

The term semantic networks encompasses a class of knowledge representation formalisms. They are made up of nodes and arcs between them. The nodes usually represent objects, concepts, or situations. The arcs represent the relations between the particular type of node. A semantic network may be viewed as an acyclic weighted graph. The relational arcs often have weights, indicating the strength of the relation associated with them. The relations are typically not two-valued, but multi-valued. CASNET[122] is an important example of an expert system which uses the semantic network knowledge representation formalism. Figure 1.4 shows the structure of a general semantic net. In the figure a partial description of the characteristics of a cat are given. The weighted links indicate aspects of a cat that are not always true.

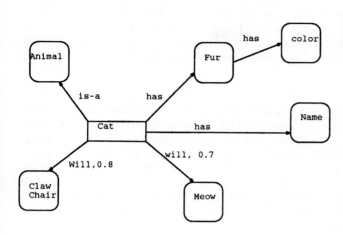

Figure 1.4. Partial Semantic Net to describe a cat.

A frame is a data structure used for representing a stereotyped situation. It is organized much like a semantic net, in many cases. It can provide built-in inheritance properties. A frame is made up of a set of slots. Slots may contain procedures, data or be pointers to other frames. We, therefore, may have nested frames. A frame system may be implemented in the context of an object-oriented knowledge representation scheme.[70] These systems, such as the common lisp object system, allow inheritance through the concept of super and subclasses. Methods, which act as procedures, will define what actions may be taken when a slot or frame object is accessed. Frames are the knowledge representation scheme employed by CENTAUR,[1] which was an improvement upon the pulmonary dysfunction expert system PUFF.[2] Frames have been combined with rules to give a powerful and flexible knowledge representation scheme. An example of a frame is given in Figure 1.5.

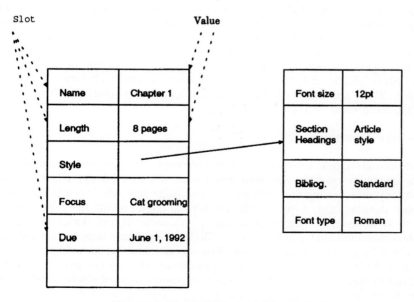

Figure 1.5. Frame example for a book chapter.

Cases are a very useful form of knowledge representation for legal and medical domains. In both of these domains, past cases are used to make decisions on present cases. Cases may be represented as monolithic textual blocks indexed by keys or in other ways beyond the scope of this chapter. They must be represented so that they are easily retrieved and it is desisable that a minimal set of cases be retrieved and that retrieval be fast.

1.1.2 The Inference Engine

The *inference* engine uses the knowledge in a particular representation to come to some expert conclusion or offer expert advice. It contains the system's general problem-solving knowledge. It is responsible for determining what piece of knowledge to use next, and scheduling other necessary actions. It will take all actions indicated, as necessary, by a piece of knowledge which is found to be true, due to the current facts presented to the expert system.

The inference engine is responsible for determining when to ask the user a question and when to search the knowledge base for the information. It must ensure that questioning is done in a concise logical manner. In most expert systems it must be responsible for dealing with imprecise and uncertain information.

An inference engine makes use of a special type of expert system knowledge called *meta-knowledge*. Meta-knowledge is knowledge about the system's knowledge. This may include such things as how best to utilize various knowledge chunks, which pieces of knowledge to use first, whether a piece of knowledge should be inferred or asked of the user, and when to stop reasoning and present a conclusion. This knowledge may be in the knowledge base in the same representation as other knowledge; it is just used in a different manner.

Inference engines operate primarily in one of two ways. They may be data driven, known as forward chaining, or they may work backwards from conclusions, known as backward chaining. A forward chaining system begins with some data and moves down the inference chain until it reaches a final node, frame or conclusion. A backward chaining system begins with some final node, frame or conclusion and works backward until it finds a complete path of evidence for one of the stopping states. MYCIN is a backward chaining expert system. XCON[87] is a forward chaining expert system. There are some expert systems which use a combination of forward and backward chaining such as SPERILL II.[94]

1.1.3 Knowledge Acquisition

We have discussed how knowledge may be represented in an expert system and how that knowledge may be used to infer some conclusions. The question that arises concerns the methodology of acquiring the knowledge. Knowledge acquisition is a bottleneck in expert system development. Typically the knowledge engineer and the expert system developer, must sit down with an expert, in many long sessions, and extract the expert's domain knowledge for use in the expert system. Alternatively, where large amounts of good data are available, knowledge may be acquired by machine learning techniques.[101, 112] At the present time the learning of rules is the best understood, most used method of automatic knowledge acquisition from data.[96, 101] Cases for case-based reasoning systems may also be acquired by learning.[103]

For interactive knowledge acquisition the knowledge engineer must have some understanding of the expert's area so that he can converse in the expert's terminology.[23] The expert will often not be able to provide general problem solving heuristics. Example problems often must be presented and the heuristics explicated in the process of the expert solving them. The initial heuristics gathered tend to be incomplete and often not quite correct. Therefore the process of gathering knowledge is incremental and coincides with the development and testing of the expert system.

In order to develop an expert system, a willing and patient expert must be found for the knowledge acquisition process. The initial sessions with the expert will be an intense process of finding the relevant information and how to get at it. Further sessions will be concerned with filling gaps in the system's knowledge and correcting any errors that may have crept in. Incorrect knowledge is acquired due to the fact that the knowledge engineer must do a lot of interpretation and weeding out in acquiring the heuristics from the expert. Also, if several experts contribute knowledge it is occasionally the case that they contradict one another to some degree. The knowledge engineer is responsible for weeding out the inconsistencies in the knowledge base.

Some tools have been developed to automate the knowledge acquisition process, such as one based on Personal Construct Theory.[19, 20] These tools are of somewhat limited utility, but can provide a good start at knowledge acquisition and in some cases take care of it completely. Also quite useful are the tools that aid the knowledge acquisition process, such as TEISESIAS[30] which aids knowledge acquisition and refinement for the expert system building tool EMYCIN.[120] It checks phrasing of clauses in rules and for completeness in the rules, as well as for contradictions. SEEK[99] gives advice about rule refinement during the development of a diagnostic-type expert system. It will help refine rules represented in the EXPERT[123] language. EXPERT is an expert system building tool. Many of the commercial expert system shells have interfaces which aid knowledge acquisition, in a manner much like what TEISESIAS.

1.2 SUITABLE PROBLEMS FOR EXPERT SYSTEMS

We will summarize the types of problems for which expert systems may currently be the solution. They are certainly a solution for problems in a narrow domain in which identifiable human experts exist and are available to aid the knowledge acquisition process. This type of problem will have little or no uncertainty involved with it. These types of problems include part of the broad class of control problems. Expert systems can be used to solve many large problems which are appropriately restricted. The restriction must rule out areas which involve common sense and analogical reasoning, as well as those that require extensive learning. An expert must be available and the complexity of the problem must not be extreme in the eyes of an experienced knowledge engineer.

Expert systems can be useful in limited domains in which there is not a human expert. These include such things as robot activities. An expert system is best applied to problems that have limited domains and well defined expertise. They are most likely to be accepted in areas that do not require extensive user interaction. As with humans, they work best in environments that do not have a large amount of errorful and uncertain information in them.

Fuzzy expert systems have been shown to work well in control domains that have imprecision in them and are to control one object. There is often tuning required in developing the fuzzy sets to represent both real-valued inputs and real-valued outputs.

Expert systems will be the solution for consultation systems of very limited domains.

They will be most accepted if they critique the expert's or practitioner's conclusions rather than coming to independent conclusions. Explanation facilities, the ease of knowledge acquisition, and making the users feel comfortable with the systems are the major obstacles for consultation systems. It is currently difficult to develop a viable expert system for a dynamic problem area.

In a problem area that appears to be suitable for development, the use of an expert system building tool to provide a prototype for one or two problems will show whether the expert system is appropriate. In the near future expert systems will be able to handle larger domains and those with a large amount of uncertainty. However, they currently are not useful in the extremes of the two above environments.

A brief discussion of the problems that currently are not amenable to expert system application will further clarify those that are. Problems that require extensive learning and re-learning are currently not possible to solve, in general, with an expert system. Domains in which reasoning by analogy is necessary tend to cause the use of case based expert systems. This is a newer knowledge representation and can be successful, but probably in only narrow restricted domains at this time. Currently the knowledge about a problem must be easily explicatable from the human expert. We do not yet know how to acquire and adequately represent common sense knowledge, although work is proceeding along the lines of building large expert systems with common sense knowledge.[83] Expert systems are not the solution for problems in the social sciences due to their great complexity and the fact that they are not well modeled. They are often not solvable by human experts and tend to involve a great amount of uncertainty. Problems in Meteorology, in which no expert is very precise, do not lend themselves well to expert system development. This indicates that as we get above a certain amount of uncertainty in our problem area an expert system is not appropriate.

Hayes-Roth[52] provides some insight into the problems which are currently beyond the scope of an expert system. We don't fully know how to reason from first principles, the core set of ideas which generally enable a skilled person to recreate many special cases. Much work in qualitative reasoning is underway.[36] We are unable to capture the nonverbal understanding of physical operations that even very young children demonstrate. For instance we know what will happen when we knock that expensive, fragile vase to the hard floor and we cringe in anticipation as it begins to fall. The effort to put this naive physics knowledge into knowledge based systems is daunting. The representation of meta-knowledge is still weak.

Most of our meta-knowledge today is involved in influencing our architectural design. Meta-representation is also lacking. That is, we lack a high level method of uniformly representing our knowledge before it is brought into the formalism appropriate for our application. This means that the wrong representation structure may be chosen, when a structure is not clearly indicated, and no easy method of recovery will exist.

1.3 UNCERTAINTY IN EXPERT SYSTEMS

It will often be the case that knowledge or information is uncertain in nature. For example there is a 40% chance of rain today in Tampa, may indicate an umbrella is a useful, but not necessary tool for the day. The representation and use of uncertain knowledge and facts is an important part of the ongoing research in expert systems. There are a number of ways to represent uncertain knowledge. Probabilistic inference networks[95] is one successful

The Evolution of Expert Systems

method. Knowledge is represented in a network where nodes of the network may be, for example, error states in a process and possible causes. The links in the network will then have conditional probabilities associated with them.

Another approach is the use of belief intervals[109, 110] in which an exact probabilistic belief is not calculated, but rather a range in which the belief may lie is calculated. These and other probabilistic methods are discussed in the International Journal of Approximate Reasoning and the annual conference on Uncertainty in AI among other sources. In this chapter, more of our focus will be on imprecision in expert systems.

1.4 ADDING IMPRECISION HANDLING TO EXPERT SYSTEMS

The imprecision that we are interested in is the kind that people effectively deal with every day. It is represented by linguistic statements such as "John has a nice car", "That is a big house", or "Check-out time is 12 noon". The first two statements are obviously imprecise. Nice, for example, might be determined by the cost of the car, the style of the car, the handling of the car, and the comfort of the car. Handling and comfort are themselves imprecise concepts. The last of the 3 example statements may seem to be crisp or precise. Yet, anyone who has traveled knows that 12:15 or earlier is certainly an acceptable check out time and the time can be moved to 1 p.m. or 1:30 p.m. with a simple phone call. Hence, the statement really is imprecise because it is taken to mean about 12.

The most successful examples of dealing with imprecision in expert systems come from the field of fuzzy control.[14, 82, 115] Fuzzy expert system controllers have been successfully used in commercial products.[26] In control (and other types of problems expert systems may be used in), the inputs are real-valued. People tend to respond to real values by labeling them with a group name. For example, 80°F might be considered warm and 90°F hot in the context of outside temperature. Real-valued attributes can present a problem for an expert system, because each value cannot be dealt with individually. Therefore, the values must be grouped together in some way. However, the boundaries of the groups are often imprecise. When does a day go from mild to warm to hot in temperature? When does a person's height move from the average to tall category? At 6 feet, 6 feet 1 inch, etc. or is it a gradual change? Fuzzy expert systems model real values for an attribute as belonging to fuzzy sets that have some overlap.

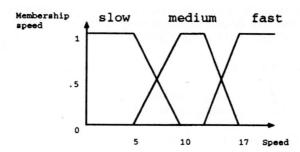

Figure 1.6. Fuzzy partition of bicycle speed.

Figure 1.6 shows 3 fuzzy sets for bicycle speed. They are slow, medium and fast and

have an overlap that is typical in fuzzy expert systems. Hence, Figure 1.6 is showing a partition of the attribute speed into 3 fuzzy sets. Given a value for speed, non-zero memberships in 1 or 2 fuzzy sets will be obtained. These values can be used as the degree of match for individual clauses or premises in the antecedents of fuzzy rules.

Figure 1.7 shows the core of a fuzzy expert system. The systems' inputs go through a fuzzifier. If the inputs are real-valued attributes they are assigned to 1 or more fuzzy sets based on the definition of the fuzzy sets stored in the knowledge base. If the attributes are nominal, they may have a fuzzy membership associated with them, but will not be modified

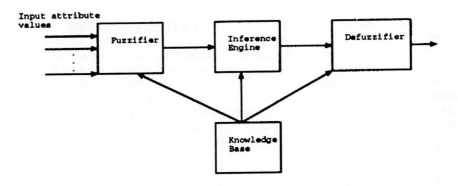

Figure 1.7. Core of fuzzy expert system.

by the fuzzifier. The inference engine works with attribute values, which have fuzzy memberships attached to them. The consequents may be fuzzy. They may be created from real-valued attributes which have been partitioned into individual fuzzy sets, such that each individual consequent contains only 1 or more of the fuzzy sets from the partition. The inference engine provides a fuzzy output, which may need to be defuzzified depending upon the application domain. Certainly, in control domains the output must be defuzzified so that there is a single well-defined control action taken.

Generally, several fuzzy rules will fire with different strengths[14, 82, 115] There are a number of different methods to determine what the output should be such as maximum of minimum firing strengths for a given consequent such as big push in controlling an object that is to be pushed. The defuzzification strategy is very important to the performance of a fuzzy expert controller and must be carefully chosen.

1.4.1 Examples of Imprecision

In the following some examples of the different types of imprecision presented above will be given. The likely source(s) of the imprecision or uncertainty in the examples will also be discussed.

It can be the case that a person is unsure of a particular piece of causal information. An example:

IF Z is tall, THEN Z played on a baskctball team. (0.8)
Z is tall (0.9).

The numeric values are certainty values associated with the rule and fact, respectively. In the above statement, it may be that enough information has been gathered to reliably assign a probabilistic uncertainty to the statement. However, this is often not the case and the value may be generated by a well-designed fuzzy membership function [48] for people who have played basketball. To determine whether a person is tall, a fuzzy membership function on heights may provide us this information. It may not be as exact as doing the work necessary to derive a reliable probability, but since fuzzy inferencing will still work reliably with small errors the correct answers can still be determined.

Another type of imprecision comes from a linguistic statement in which the boundary is not clearly defined. For example:

Mary is middle-aged
IF the inventory is high, THEN production should be slowed

The terms *middle-aged*, *high*, and *slowed* are imprecise terms. In the context of fuzzy set theory, they are called fuzzy terms. This type of information is modeled well by linguistic fuzzy sets. It is true that they could be modeled, for example, with the Dempster-Shafer theory of evidence, but it is our contention that fuzzy sets provide a more intuitive method and will provide accurate answers.

Both uncertainty and imprecision may exist in some statements. For instance:

IF the wind is high THEN the sailing should be good (0.8)

In this case, the fuzzy information is contained in the terms high and good. The overall statement only holds most (0.8) of the time.

The uncertainty could also be fuzzy. In the statement,

Ed is masterful sailor, (0.8 - 0.9)

a fuzzy range with which it may be believed is given. The range could be interpreted as *about 0.85*.

It is not our intention to cover anything other than how fuzzy set theory may be used in expert systems. There are places where the uncertainty may be probabilistic in nature and probability based theories will provide the best models. Often it will be the case that fuzzy sets may be used, possibly in conjunction with probability, to model the uncertainty and imprecision inherent in real world problems.

There is imprecision in most areas of expert systems. Most of the interesting domains to which an expert system may be applied have some significant amount of imprecision in them. It should be noted that some of the applications with the least amount of imprecision in them have been most successfully attacked. Those with large amounts of imprecision have not been as successfully solved and some are beyond the current reach of expert systems. This indicates that uncertainty and imprecision are very important aspects of an expert system. They must be handled in a sound manner both theoretically and practically.

1.4.2 Some Examples of Fuzzy Expert Systems

The theory of fuzzy sets and fuzzy logic is well-founded and strong. The theories have been in existence for over 25 years and have been used in many control applications especially in Japan. Fuzzy logic is employed in the control of trains in Japan.[91] It has been very successful and an expert fuzzy controller is on-line in the city of Sendai, Japan. It provides smoother changes in velocity than non-fuzzy systems have yet achieved. Fuzzy set theory is used in the cement kiln controller of Mamdani.[85] In both of these cases linguistic fuzzy variables are used to do internal calculation and provide the control strategy. Fuzzy control has also been used in video camera stabilization, vacuum cleaners, automatic transmissions[107] and other control applications.

Fuzzy expert systems properly used will allow fuzzy reasoning schemes to be developed and applied to a wide range of problems without constant minor changes. A unified theoretically sound set of methods can be developed for reasoning under uncertainty in expert systems.

Whalen and Schott[125] use a fuzzy linguistic logic system with production rules to suggest appropriate forecasting techniques for sales predictions in their expert system. They use possibilities and work in a backward chaining manner. They start with the possibilities of all consequents equal to one and attempt to reduce the possibilities until they find an irreducible set. SPERILL-II,[94] is an expert system for damage assessment of existing structures, uses fuzzy sets to represent imprecise data. FLOPS[25] is a fuzzy rule based production expert system. Input to the system is a vector all of whose components are fuzzy sets. As output it produces a fuzzy set of consequents. Incoming information is pattern matched against rules to provide a fuzzy set of fireable rules, which are then executed. They continue this process until they can no longer find rules to fire and then produce a set of fuzzy consequents. FLOPS was sold commercially for years. It has been successfully used to model several different expert domains.[25]

The Z-II system[84] is a fuzzy expert system shell, which effectively deals with both uncertainty and imprecision. It has been used to construct several expert systems. The domains have been medical diagnosis, psychoanalysis and risk analysis. It has been found that the experts felt it allowed them to express their information in a natural manner. It allows knowledge to be expressed in fuzzy linguistic terms. An example rule is the following:

IF your interest in analyzing the human body is high
THEN your overall interest in medicine should be high (0.95)

As long as the linguistic terms are defined by some reasonable membership function, this is a very effective knowledge representation format for imprecise and uncertain information.

Another important type of system for which fuzzy sets provide a powerful basis is common sense reasoning. This area has been researched somewhat less than the use of fuzzy expert systems for control. The theory of usuality put forth by Zadeh[132] provides a tool to incorporate some common sense information into expert systems. The concept of *usuality* relates to events which are usually true or have a high probability of occurrence. A usuality qualified proposition may have an implicit *usually*. It is then called a *disposition*. Some common dispositions are snow is white, *Florida is a warm state*, and *windy days are good for sailing*. Now usually may be interpreted as a fuzzy quantifier, which is basically a fuzzy proportion based on an implicit or explicit sigma count. The theory of usuality may be applied to ordinary situations and provide a method to effectively represent knowledge about events or items which are often true. This includes many common sense concepts.

Possibly the biggest weakness of non-fuzzy methods of dealing with imprecision and uncertainty is their handling of linguistic terms. Fuzzy set theory provides a natural method for dealing with the linguistic terms by which an expert will describe a domain. An imprecise numeric term can be effectively described by a fuzzy number.[67] Other terms are simply mapped to and from fuzzy sets. The use of fuzzy set theory in expert systems has caused an evolution of them to many control problems and the opportunity to address a further set of problems in concert with other uncertainty handling methods.

1.5 THE EVOLUTION TO THE HYBRID CONCEPT

In previous sections we have studied the evolution from expert systems to fuzzy expert systems. It is clear from recent research[68,69] that the ability of intelligent autonomous systems to learn in an uncertain environment is enhanced by incorporating event-driven artificial neural network learning mechanisms into fuzzy expert systems technology. Those learning techniques may enable the fuzzy expert system to modify and enrich its knowledge structures autonomously.

It is the purpose of this section to outline the evolution from fuzzy expert systems to hybrid systems which utilize both an expert system or a fuzzy expert system and an event-driven artificial neural network. The overall system acquires knowledge from the fuzzy expert system and is capable of learning and acquiring new knowledge and refining existing knowledge via the neural network by utilizing associative learning techniques.

Machine learning in an uncertain or unknown environment is of vital interest to those working with intelligent systems. The ability to garner new information, process it, and increase the understanding/capability of the machine is crucial to the performance of autonomous systems. Among others, the field of artificial intelligence provides two approaches to the problem of knowledge engineering—expert systems and neural networks. Harnessing the power of these two techniques in a hybrid, cooperating system holds great promise.

Expert systems and fuzzy expert systems[67-69] are strongly tied to knowledge-based techniques for gathering and processing information. Knowledge representation in such systems is most often in the form of rules garnered through consultation with human experts. Coupling the methods of approximate reasoning with knowledge-based techniques yields systems which model human decision-making. There are many examples of rule based systems which function as experts in a given domain, e.g., trouble-shooting for complex mechanical processes, medical diagnosis systems, and financial risk assessment. Expert systems provide a ready mechanism for explanation why certain decisions are made, even when the human expert is unable to articulate the chain of reasoning leading to a decision. This trace of the reasoning process is often crucial to those maintaining the system.

A major disadvantage of knowledge-based systems is their reliance upon consultation with human experts for new information. Furthermore, autonomous learning in an expert system, which is rare, does not usually include the capability to synthesize new knowledge but is limited instead to dependence upon structures the designer builds in to asses the similarity between situations or to generalize upon sets of similar rules.

Neural networks are data-driven systems based on an architecture of many simple processing units which are interconnected. The knowledge of a neural net resides in the connections between these processing units and in the strengths of the connections. Neural networks are especially applicable to problems which involve large numbers of weak constraints. They have been successfully applied to perceptual tasks such as pattern recognition, vision processing, and speech synthesis.[57, 106]

The ability to gracefully handle minor inconsistencies or conflicts in the data is an advantage that neural network systems hold over most expert systems. A robust intelligent system must be able to handle conflicting information from different experts, or some degree of contamination in incoming data, without too much degradation in performance.

There are many scenarios in which both types of reasoning, knowledge-based and data-driven, are appropriate. Harnessing the power of both expert systems and neural networks in a system which allows for imprecise information and/or uncertain environments would yield a system more powerful than either system standing alone.

Several preliminary investigations of such a hybrid system have been made. Gallant[41] proposed a "connectionist expert system" which is an expert system with a neural network for its knowledge base. It allowed him to automate the construction of the knowledge base using examples, thereby reducing the development effort for the expert system. Fu[38] has developed a "conceptualization network" from a rule-base and inference engine of an expert system which can recognize and assign blame for incorrect rules. Hudson, Cohen and Anderson[62] describe a system for medical diagnosis which uses a neural network to categorize situations directly from the data and then feed a reduced set of information into an expert system.

The preliminary nature of this field as well as its promise is evident in these as well as related investigation.[4, 6, 22, 43-46, 69, 77, 86-88, 114-118, 134] The problems considered involved the management of large pools of data and refining the process of passing pertinent data on to the expert system.

1.6 HYBRID INTELLIGENT SYSTEMS

Learning is an essential component for any intelligent system. Michalski et al.,[89] summarize the classic strategies and orientations in machine learning. An autonomous system should be able to learn in an unsupervised situation by experimentation, classification, recognition or similarity, generalizing and applying appropriate previous solutions or hypothesizing new solutions to situations never before encountered by the system.

Learning in a neural network without the benefit of an initial base of knowledge can be very slow to converge. Therefore, the premise in this paper is that learning can be implemented more efficiently in the neural network when the expert system supplies the meta-knowledge to begin the learning process as well as accumulated knowledge in the system.[49-60, 92, 128]

We are currently evaluating several models for the hybrid system. As a general premise, we believe that the most effective model is one in which the expert system begins with a base of knowledge which is necessarily incomplete, a neural network layer takes the knowledge from the expert system and modifies it through learning, and all information can be passed easily and transparently from one part of the system to another as needed.

One may consider several configurations of this basic model which have different uses:

1. Everything learned in the neural network is passed back to the expert system. In effect, the neural network is training the expert system. In this version, the user of the system is always able to trace the decision-making process via the expert system.

The Evolution of Expert Systems

2. When a problem is presented to the hybrid system, it is partitioned into segments which are evaluated to be appropriate for solution by either the expert system layer or by the neural network layer. The solution to the problem is a hybrid of the segment solutions. In this version, the two layers act as cooperating partners, each doing what it does best, keeping functional overlap to a minimum.

3. An entire network of smaller systems, expert systems and neural networks, cooperate and communicate to learn in different modes, or in different domains. Each part is designed with a different part of the problem solution process in mind.

The fluid transfer of information from one type of system to the other obviously is crucial. Validation of the learning process and verification that new information has been correctly communicated and incorporated is also of interest. As Kosko[77] points out, it is important to consider the stability of unsupervised learning. The robustness of such systems and their behavior in the face of perturbations to the learning environment are open questions.

Another issue of interest is the application of uncertainty management techniques within the hybrid intelligent system to better model the human reasoning process. Most conventional rule-based systems allow the use of certainty factors to represent the fact that a rule does not hold true for all situations satisfying the antecedent conditions. In fuzzy reasoning, we consider the degree to which a rule agrees with our current understanding of reality rather than the probability that it is a true description of that reality.[35-38, 51, 61-64, 90, 92, 97-98, 111]

The sensitivity of learning in the neural network to the use of linguistic variables as weights and linguistic hedges will be investigated. One way to measure this sensitivity is to compare the rate of change in what has been learned against the change in the certainty factors or weights.

It should be pointed out that the transfer of knowledge between system components is bidirectional and it is precisely the learning capabilities of neural networks that enable the intelligent system to infer new rules or modify existing rules based on neural network performance. The division of labor, by providing the system with whatever knowledge is available a priori through the expert system and the knowledge-base, and by developing optional learning strategies for the neural network, is precisely the technology that will provide us with fast, autonomous effective learning on top of previously acquired knowledge.

It is the cooperation of an expert system with a **given** knowledge-base and a neural network with the **learning capabilities** that enables this technology to execute tasks in an autonomous, imprecise and somewhat unpredictable environment.

1.7 APPLICATIONS TO INTELLIGENT CONTROL

Conventional control systems design methodology involves the construction of a mathematical model describing the dynamic system to be controlled and the application of analytical techniques to this model to derive a control law. Robust control design produces a constant gain controller which stabilizes a class of linear systems over a range of system parameters. Adaptive control adjusts the controller characteristics to stabilize a system with unknown parameters.

1.7.1 Autonomous Control

Adaptive Control is particularly critical for enhanced functionality and error detection and correction. Error detection and correction are associated with monitoring-execution mechanisms to ensure compliance with the expected task sequence, to measure expected errors, and to correct them dynamically.

Conventional control techniques break down, however, when a representative model is difficult to obtain due to uncertainty, or sheer complexity. Model uncertainty is a serious problem in designing intelligent robot control laws. Often it is impossible to adequately represent system characteristics such as nonlinearity, time delay, saturation, time-varying parameters, and overall complexity. Thus, autonomous intelligent control systems require significantly enhanced capabilities to achieve real-time operational responses when the decision-making process is based on incomplete information, uncertainty, and competing constraints.

Biological systems, on the other hand, handle ill-structured problems with remarkable ease and flexibility. They are quite successful in dealing with uncertainty, complexity, and nonlinearities. They coordinate smoothly many degrees of freedom during the execution of dexterous manipulative tasks within unstructured environments, solve complex planning problems with apparent ease, and are able to adapt their structure and function.

Important features of biological control systems include:

1. Hierarchical and modular processing architecture.
2. Distributed computation among the various levels of the hierarchy.
3. Utilization of tightly integrated, yet distinct, forms of sensorimotor processing during the acquisition of motor skills.

It seems therefore desirable to turn to biologically-inspired paradigms in developing efficient processing architecture and learning procedures to improve the performance and adaptability of intelligent control systems. These biological paradigms, namely, artificial neural networks, seem to be potentially useful in treating many problems that cannot be handled by traditional analytical approaches. For example, back-propagation neural networks currently are the most prevalent neural network architectures for control applications because they have the capability to "learn" system characteristics through nonlinear mappings.

1.7.2 The Combined Approach

Current adaptive control techniques reveal fundamental shortcomings in terms of implementing robot control laws. Adaptive control laws, such as Model Reference Adaptive Control, Self Tuning Regulator, and Gain Scheduling[5, 81] are nonlinear control laws which are difficult to derive, their complexity grows geometrically with the number of unknown parameters, they are not robust, they are conditionally stable, and often they are not suitable for real-time applications.

In contrast, control architectures based on neural networks are specifically suitable to implement general purpose trainable adaptive controllers for robotic control. Trainable adaptive controllers are process controllers where much of the design is done online via training rather than programming.

Neural networks are inherently robust and are massively parallel, adaptive, dynamical systems modeled on the general features of biological networks. Due to the availability of advanced VLSI implementation techniques and the demand for massive parallelism to achieve real-time information processing, there has been tremendous interest in the applications of neural networks to achieve human-like performance in the field of robotics.

Neural networks interact with objects of the real world and their statistical characteristics in much the same way living beings do. They consist of densely interconnected processing elements, or neurons. Each neuron is provided with the ability to self-adjust some of the coefficients in its governing differential equations. Thus, the network as a whole becomes a self-adapting dynamic system, capable of learning and self-organizing, and operating in a highly parallel distributed manner, most suitable for high-performance information processing.

Collectively, neurons with simple properties, interacting according to simple rules, can accomplish complex functions such as generalization, error correction, information reconstruction, pattern analysis, and learning. Their paradigmatic strength for potential applications, which require solving intractable computational problems or adaptive modeling, arises from their ability to achieve functional synthesis, and thereby learn topological mappings and abstract spatial, functional, or temporal invariances of these mappings. Thus, relationships among multiple continuous-valued inputs and outputs can be established, based on presentation of various representative examples.

Once the underlying invariances have been learned and encoded in the topology and the interconnections weights, the neural network can generalize to solve arbitrary problem instances. Since the topological mappings for problem-solving are acquired from real-world examples, the functionality of the neural network is not limited by assumptions regarding parametric or environmental uncertainty. Thus, neural networks provide an attractive algorithmic basis for solving fundamental design problems of autonomous intelligent control systems.

In addition to that, neural networks also provide a greater degree of robustness or fault tolerance than the conventional von Neumann sequential machines. Damage to a few neurons or connections does not impair overall performance significantly. Since most neural network models tend to adapt connection weights so as to self-organize internal representations in response to the continuously changing inputs, adaptation also provides a degree of robustness by compensating for minor variabilities in the characteristics of neurons.

The unique computational properties of neural networks have already been used successfully in a number of applications including associative retrieval of information,[3, 57] optimization[60, 116] signal regularity extraction[271] perceptual inference[55] feature discovery[106] sensor motor control,[471] identification and control of dynamical systems,[93] and robot control.[50, 108]

The benefits of the combined approach are:

1. Realization of fast decision making and control since computations are done in parallel.

2. Fast adaptation to a large number of parameters since the convergence rate of a neural network to a steady-state is independent of the number of neurons [58], [60].

3. Adaptation to parameter variations over continuous and discrete time domains since the neural network may be continuous or discrete.

4. The control laws do not have to be stated explicitly since learning can be done through examples.[27]

5. Fault tolerance due to the distributed representation of information in neural networks which provides graceful degradation.

6. Robustness to unmodeled parameters and uncertainty since the neural network is capable of generalizations.

1.8 CONCLUSION

The hybrid intelligent system is the next step for the approach presented in the previous sections. It provides a integrated tool which could form the basis for a potentially fruitful approach to intelligent robot control problems. The expert system provides us with a tool to handle a priori knowledge whereas the neural network offers potentially powerful collective-computation techniques, as well as learning capabilities in an adaptive environment.

We see the use of hybrid intelligent system in robot control as a natural step in the evolution of robot control methodology to meet new challenges. The cooperative structure of the hybrid system, incorporating a priori knowledge, learning capabilities, and massive parallelism, offers solutions to several critical issues that are essential for designing intelligent systems.

Knowledge based systems provide a convenient mechanism for automated complex decision-making with task-specific knowledge being defined explicitly. Neural networks, on the other hand, encode knowledge implicitly, adjusting internal weights so that their input/output relationships remain consistent with observed training data.

An important emphasis is on the mechanism for shifting knowledge and control between the two components of the hybrid system, in order to utilize the strengths of each processing technique. First, the knowledge-based system determines how to accomplish a given control objective using rules and algorithms within the knowledge base. It then teaches the neural network how to accomplish the same task by having the neural network observe and generalize on knowledge-based task execution. Based on the performance of the neural network, knowledge is transferred back to the knowledge-based system to infer new rules or modify existing ones if applicable.

One scheme proposed to classify memory and learning distinguishes between declarative and reflexive mechanisms.[80] Motions involving declarative mechanisms are characterized by inference, comparison and evaluation, and provide insight into how something is done and why it is done. Motions involving reflexive mechanisms relate specific responses to specific stimuli, are automatic, and require little or no thought. Tasks initially learned declaratively often become reflexive through repetition. Conversely, when familiar tasks are attempted in novel situations, reflexive knowledge must be converted back into declarative form to become useful. This shifting of task-specific knowledge between declarative and reflexive forms plays a fundamental role in skill acquisition.

In terms of this scheme, the declarative form of processing is implemented in our hybrid intelligent system by knowledge-based expert systems whereas the reflexive form of processing is implemented using neural networks.

The integration of the fuzzy expert system and neural network would minimize the "learning time" through the use of the expert system for a priori knowledge, as well as utilizing the learning capabilities and the parallelism of the neural network. The use of the hybrid system to achieve just these two modifications could serve to significantly advance the present capabilities or robotic control systems. The learning capabilities of the system are one of the main strengths of the hybrid approach. Integrated with expert system technology it could be used as a powerful tool for addressing robotic needs for adaptation to both task and environment changes, selection of optional task features, and incorporating a priori knowledge regarding uncertain environments.

ACKNOWLEDGMENTS

Hall's research partially supported by a grant from the Flodida High Technology and Industry Council, CIM section and a grant from the Whitaker Foundation, and the National Cancer Institute (CA59 425-01). Kandel's research partially supported by a grant from the Council on Research and Creativity, USF.

REFERENCES

1. Aikens, J.S. "Prototypical Knowledge for Expert Systems." *Artificial Intelligence*. Vol. 20 (1983), pp. 163-210.

2. Aikens, J.S., J.C. Kunz, and E.H. Shortliffe. "PUFF: An Expert System for Interpretation of pulmonary Function Data." *Computers and Biomedical Research*. Vol. 16 (1983), pp. 199-208.

3. Anderson, J. A. "Cognitive and psychological computation with neural models." *IEEE Trans. Syst., Man. and Cybernetics*. Vol. SMC-13, (1983), pp. 799-814.

4. Anderson, J. R. "A mean field computational model for PDP." *Proc. 1988 Connectionist Models Summer School*. D. Touretzky, G. Hinton, and T. Sejnowski, Eds., San Mateo, CA: Morgan Kaufmann (1989), pp. 217-223.

5. Astrom, K. J. "Adaptive Feedback Control." *Proc. IEEE*. Vol. 75 (1987), pp. 185-217.

6. Baldi, P. and K. Hornik. "Neural networks and principal component analysis: Learning from examples without local minima." *Neural Networks*. Vol. 2 (1989), pp. 53-58.

7. Bandler, W. and L.J. Kohout. "Activity Structures and Their Protection." *Proceedings 1979 International Meeting of the Society for General Systems Research*. Louisville, Kentucky (1979), pp. 239-244.

8. Bandler, W. "Representation and Manipulation of Knowledge in Fuzzy Expert Systems." Presented at Workshop on Fuzzy Sets and Knowledge-Based Systems, Queen Mary College, University of London, England, UK (1983).

9. Barnden, J. and K. Srinivas. "Dissolving variables in connectionist combinatory logic." *Proc. Int. Joint Conference on Neural Networks*. San Diego, CA (June 17-21), Vol. III (1990), pp. 709-714.

10. Barr, A. and E.A. Feigenbaum. *The Handbook of Artificial Intelligence I.* William Kaufmann, Los Altos, CA (1981).

11. Baum, E. B. "What size net gives valid generalization?" *Neural Computation*, Vol. 1, No. 1, pp. 151-160.

12. Baum, E. B. and G. Wilczek. "Supervised learning of probability distributions by neural networks." D. Z. Anderson, Ed. *Neural Information Processing Systems.* American Institute of Physics(1988), pp. 52-61.

13. Bennett, J.S., R. S. Engelmore. "Experience Using EMYCIN." B. Buchanan and E. Shortliffe, Eds. *Rule-Based Expert Systems.* Addison-Wesley, Reading, MA.(1984), pp. 314-328.

14. Berenji, H.R. and P. Khedkar. "Learning and tuning fuzzy logic controllers through reinforcements." *IEEE Transactions on Neural Networks*, 3 (5), (1992).

15. Bigufi, J. and K. Goolsbey. "Integrating neural networks and knowledge-based systems in a commercial environment." *Proc. Int. Joint Conference on Neural Networks*, Washington, D.C., (June 18-22), Vol. II (1989), pp. 463-466.

16. Bilbro, R., T. K. Mann, W.E. Miller, D.E. Snyder, Van den Bout and M. White. "Optimization by mean field annealing." D. S. Touretzky, Ed. *Advances in Neural Information Processsing Systems.* Vol. I, San Mateo, CA: Morgan Kaufmann (1989), pp. 91-98.

17. Bochereau, L. and P. Bourgine. "Rule extraction and validity domain on a multilayer neural network." *Proc. Int. Joint Conference on Neural Networks*, San Diego, CA, (June 17-21), Vol. 1 (1990), pp. 97-100.

18. Bonissone, P.P. and A.L. Brown Jr. "Expanding the Horizons of Expert Systems." *Proc. of the Conference on Expert Systems and Knowledge Engineering.* Gottlieb Duttwailer Institute, Zurich, Switzerland (April, 1985).

19. Boose, J.H. "Personal Construct Theory and the Transfer of Human Expertise." *Proceedings of the National Conference on Artificial Intelligence,* Austin, TX (1984), pp. 27-32.

20. Boose, J.H. "A survey of knowledge acquisition techniques and tools." *Knowledge Acquisition*, 1 (1), (1989), pp. 3-37.

21. Bounds, D. G., P. J. Lloyd, B. Mathew, and G. Waddell. "A multilayer perceptron network for the diagnosis of low back pain." *Proc. IEEE Int. Conf. Neural Networks*, Vol. II (1988), pp. 481-489.

22. Bradshaw, G. R., R. Fozzard, and L. Ceci. "A connectionist expert system that actually works." D. S. Touretzky, Ed. *Advances in Neural Information Processing Systems.* Vol. I, San Mateo, CA: Morgan Kaufman (1989), pp. 248-255.

23. Buchanan, B.G. and D.C. Wilkins. *Knowledge Acquisition and Learning*. Morgan Kaufmann, San Mateo, CA (1992).

24. Buckley, J.J., W. Siler, and D. Tucker. "A fuzzy expert system." *Fuzzy Sets and Systems*, V. 20, No. 1 (1986), pp. 1-16.

25. Buckley, J.J, Siler, W. and D. Tucker. "FLOPS, A Fuzzy Expert System: Applications and Perspectives." C.V. Negoita and H. Prade, eds. *Fuzzy Logics in Knowledge Engineering*, Verlag TUV Rheinland, Germany (1986).

26. "Software that can dethrone computer tyranny." *Business Week* (April 6, 1992), pg. 92.

27. Carpenter, G. A. and S. Grossberg. "A massively parallel architecture for a self-organizing neural pattern recognition machine." *Computer Vision, Graphics, and Image Processing*. Vol. 37 (1987), pp. 54-115.

28. Corkhill, D.D., K.Q. Gallagher, and P.M. Johnson. "Achieving flexibility, efficiency and generality in blackboard architectures." *Proceedings of AAAI-87*(1987).

29. DARPA. *Neural Network Study*, Fairfax, VA: AFCEA International Press (1988).

30. Davis, R. and D.B. Lenat. *Knowledge-Based Systems in A.I.*, McGraw-Hill, NY (1982).

31. Dempster, A.P. "Upper and Lower Probabilities Induced by a Multivalued Mapping." *Ann. Math. Stat.*, Vol. 38 (1967), pp. 325-339.

32. Deprit, E. "Implementing recurrent back-propagation on the connection machine." *Neural Networks*, Vol. 2 (1989), pp. 295-314.

33. Erman, L.D., R.D. Fennel, and D.R. Reddy. "System Organizations for Speech Understanding: Implications for Network and Multiprocessor Computer Architectures for A.I." *IEEE Transactions on Computers*, Vol. C-25, No. 4 (1976), pp. 414-421.

34. Erman, L.D. "The HEARSAY-II Speech-understanding System: Integrating Knowledge to Resolve Uncertainties." *Computing Surveys*, Vol. 12, No. 2 (1980), pp. 213-253.

35. Fahlman, S. E. "Faster learning variations on back propagation: An empirical study." D. Touretzky, G. Hinton and T. Sejnowski, Eds. *Proc. 1988 Connectionist Models Summer School*. San Mateo, CA: Morgan Kaufman (1989), pp. 38-51.

36. Forbus, K. "Qualitative Process Theory." *Artificial Intelligence*, V. 24 (1984).

37. Friedman, M., M. Schneider, and A. Kandel. "The Use of Weighted Fuzzy Expected Value (WFEV) in Fuzzy Expert Systems." *Fuzzy Sets and Systems*. Vol. 31 (1989), pp. 37-45.

38. Fu, L. "Integration of Neural Heuristics into Knowledge-based Inference." *Connection Science*. Vol. 1 (1989), No. 3.

39. *Fuzzy CLIPS Version 2.1 User's Guide.* Knowledge Systems Laboratory, Institute for Information Technology, National Research Council Canada, Ottawa, Ontario, Canada KlA OR6.

40. Gaines, B.R. "Foundations of Fuzzy Reasoning." Gupta *et al.*, eds. *In Fuzzy Automata and Decision Processes.* North-Holland, NY (1977), pp. 19-76.

41. Gallant, S. I. "Connectionist expert system." *Comm.*, ACM, Vol. 31, No. 2 (1988), pp. 152-169.

42. Gaschnig, J. "PROSPECTOR: An Expert system For Mineral Exploration." D. Michie, ed. *In Introductory Readings in Expert Systems.* Science Publishers Inc., NY (1982).

43. Giarratano, J. and G. Riley. *Expert Systems: Principles and Programs.* Boston, MA: PWS-Kent Publishing Co. (1989).

44. Glover, D. E. "An optical Fourier/electronic neuro-computer automated inspection system." *Proc. IEEE Int. Conf. Neural Networks.* Piscataway, NJ: IEEE, Vol. I (1988), pp. 569-576.

45. Graf, H. P., et al. "VLSI implementation of a neural network memory with several hundreds of neurons." J. S. Denlcer, ed. *Neural Networks for Computing.* Snowbird, UT, 1986, AIP Conf. Proc. 151, New York: American Institute of Physics (1986), pp. 182-187.

46. Graf, H. P. and L. D. Jackel. "Analog electronic neural network circuits." *IEEE Circuits and Devices Mag.* (July, 1989), pp. 44-49.

47. Grossberg, S. and M. Kuperstein. *Neural Dynamics and Adaptive Sensory Motor Control.* North-Holland (1986).

48. Hall, L.O. and A. Kandel. *Designing Fuzzy Expert Systems.* Verlag TUV Rheinland, Germany (1986).

49. Hall, L. and S. Romaniuk. "FUZZNET toward a fuzzy connectionist expert system development tool." *Proc. Int. Joint Conference on Neural Networks.* Washington, D.C., (Jan 15-19, 1990), Vol II, pp. 483-486.

50. Handelman, D. A., S. H. Lane, and J. J. Gelfand. "Integrating neural networks and knowledge-based systems for intelligent robotic control." *IEEE Control Systems Magazine.* Vol. 10 (1990), pp. 77-87.

51. Handelman, D., et al. "Integration of knowledge-based system and neural network techniques for autonomous learning machines." *Proc. Int. Joint Conference on Neural Networks.* Washington, D.C., (June 18-22, 1989), Vol. I, pp. 683-688.

52. Hayes-Roth, F. "The Knowledge-Based Expert System: A tutorial, Computer." 17, No. 9 (1984), pp. 11-28.

53. Hecht-Nielsen, R. "Neurocomputing: Picking the human brain." *IEEE Spectrum.* Vol. 25, No. 3 (1988), pp. 36-41.

54. Hinton, G. E. and J.A. Anderson, eds. *Parallel Models of Associative Memory.* Hilsdale, NJ: Erlbaum (1981).

55. Hinton, G. E. and T. Sejnowski. "Optimal perceptual inference." *Proc. IEEE Conf. on Computer Vision and Pattern Recognition* (1983), pp. 448-453.

56. Holyoak, K. J. "Symbolic connectionism: A paradigm for third-generation theories of expertise." Presented at Study of Erpertise: Prospects and Limits Conf., held at the Max Planck Institute Berlin, (June, 1989).

57. Hopfield, J. J. "Neural networks and physical systems with emergent collective computational abilities." *Proc. Nat. Acad. Sci.*, Vol. 79 (1992), pp. 2554-2558.

58. Hopfield, J. J. "Neurons with graded response have collective computational properties like those of two-state neurons." *Proc. Nat. Acad. Sci.* Vol. 81 (1984), pp. 3088-3092.

59. Hopfield, J. J. and D. W. Tank. "Neural computation of decisions in optimization problems." *Biological Cybernetics.* Vol. 52 (1985), pp. 141-152.

60. Hopfield, J. J. and D. W. Tank. "Neural computation of decisions in optimization problems." *Biological Cybernetics.* Vol. 52 (1985), pp. 1-12.

61. Huang, W. Y. and R.P. Lippmann. "Comparisons between neural net and conventional classifiers." M. Claudill and C. Butler, Eds. *Proc. IEEE Fisrt Int. Conf. Neural Networks.* Piscataway NJ: IEEE, Vol. IV (1987), pp. 485-492.

62. Hudson, P. L., M.E. Cohen and M.F. Anderson. "Use of neural network techniques in a medical expert system." *Proc. 3rd Congress of the International Fuzzy Systems Association* (1989).

63. Jagota, A. and O. Jakubowicz. "Knowledge representation in a multilayered Hopfield network." *Proc. Int. Joint Conference on Neural Networks.* Washington, D.C., June 18-22, Vol. I (1989), pp. 435-442.

64. Jamison T. A. and R. J. Schalkoff. "Image labeling: A neural network approach." *Image and Vision Computing.* Vol. 6, No. 4 (1988), pp. 203-213.

65. Kacprzyk, J. and R. Yager. "Emergency-oriented Expert Systems: A Fuzzy Approach." *Tech. Report MII-213/247.* Machine Intelligence Institute, Iona College, New Rochelle, NY (1982).

66. Kandel, A., ed. *Special Issue on Expert Systems, Information Sciences.* Vol. 37, Nos. 1-3 (Dec. 1985).

67. Kandel, A. "Fuzzy Mathematical Techniques with Applications." Addison-Wesley, Reading, MA (1986).

68. Kandel, A., ed. *Fuzzy Expert Systems.* CRC Press, Boca Raton, FL (1992).

69. Kandel, A. and G. Langholz, ed. *Hybrid Architectures For Intelligent Systems.* CRC Press, Boca Raton, FL (1992).

70. Keene, S.E. *Object-oriented Programming in Common Lisp.* Addison-Wesley, Reading, MA (1988).

71. Koch, C., J. Lou, C. Mead, and J. Hutchinson. "Computing motion using resistive networks." *Proc. SPIE Int. Soc. Opt. Eng.*(1988), pp. 108-113; also J. Hutchinson, C. Koch, J. Luo and C. Mead. "Computing motion using analog and binary resistive networks." *Computer* (1988), pp. 52-63.

72. Koch, C. J. Marroquin, and A. Yuille. "Analog neuronal networks in early vision." *Proc. Nat. Acad. Sci.* Vol. 83 (1986), pp. 4263-4267.

73. Kohonen, T. "Self-organized formation of topologically correct feature maps." *Biological Cybernetics.* Vol. 43 (1992), pp. 59-69.

74. Kohonen, T. "Self-Organization and Associative Memory." Springer-Verlag, Berlin (1984).

75. Kolodoner, J.L. "Extending problem solver capabilities through case-based inference." *Proceedings Fourth International Conference on Machine Learning.* Morgan-Kaufmann Publishers (1987), pp. 167-178.

76. Kolodoner, J.L. "Maintaining organization in a dynamic long-term memory." *Cognitive Science*, 7 (1983), pp. 243-280.

77. Koslso, B. "Unsupervised Learning in Noise." *Proc. Int. Joint Conf. on Neural Networks* (1989).

78. Kuczewski, R. M., M. H. Myers, and W. J. Crawford. "Exploration of backward error propagation as a self-organizational structure." M. Caudill and C. Butler, Eds. *Proc. IEEE First Int. Conf. Neural Networks.* Vol. II, Piscataway, NJ: IEEE (1987), pp. 89-95.

79. Kulp, R. L. and A. S. Gilbert. "Combination of dissimilar features in neural networks for ship recognition." *Johns Hopkins Univ. Appl. Phys. Lab. Rep.* FIA (1)-89U-022, (October, 1989).

80. Kupfermann. "Learning." A. Kandel and J. Schwartz, Eds. *Principles of Neural Science.* Elsevier (1985).

81. Landau, Y. D. *Adaptive Control: The Model Reference Adaptive Control Approach.* Marcel Dekker (1979).

82. Lee, C.C. "Fuzzy Logic in Control Systems: Fuzzy Logic Controller-Part I and Part II." *IEEE Transactions on Systems, Man and Cybernetics.* V. 20, N. 2 (1990), pp. 404-435.

83. Lenat, D.B. and R.V. Guha. "Building Large Knowledge-Based Systems." Addison-Wesley, Reading, MA (1990).

84. Leung, K.S. and W. Lam. "Fuzzy Concepts in Expert Systems." *Computer* (Sept. 1988), pp. 43-56.

85. Mamdani, E.H. "Fuzzy Logic Controllers with Industrial Applications." *Proc. JACC.* San Francisco, CA (1980).

86. McClelland, J. L. and D. E. Rumelhart. "Distributed memory and the representation of general and specific information." *Journal of Experimental Psychology.* 114, No. 2 (1985), pp. 159-188.

87. McDermott, J. *Domain Knowledge and the Design Process, Design Studies.* Vol. 3 (1982), No. 1.

88. Mead, C. *Analog VLSI and Neural Systems.* New York: Addison-Wesley (1989).

89. Michalski, R. S., J. G. Carbonell, and T. M. Mitchell. *Machine Learning.* Volumes 1 and 2, Morgan Kaufman, Palo Alto, CA (1983, 1986).

90. Mlller, M. I., B. Roysam, and K. R. Smith. "Mapping rule-based and stochastic constraints to connection architectures: Implication for hierarchical image processing." *Proc. SPIE Int. Soc. Opt Eng.* (1988), pp. 1078-1085.

91. Miyamoto, S. and S. Yasunobu. *Predictive Fuzzy Control and its Application to Automatic Train Operation Systems.* Fuzzy Information Processing Society Conference (1984).

92. Murphy, J. H. "Probability-based neural networks." *Proc. International Joint Conference on Neural Networks*, Washington, D.C., (Jan 15-19, 1990), Vol. I, pp. 451-454.

93. Narendra, K. S. and K. Parthasarathy. "Identification and control of dynamical systems using neural networks." *IEEE Trans. Neural Networks.* Vol. 1 (1990), pp. 4-27.

94. Ogawa, H., K.S. Fu, and J.P.T. Yao. "Knowledge Representation, and Inference Control of SPERILL-II." ACM Conference Proceedings, (November, 1984).

95. Pearl, J. *Probabilistic Reasoning in Intelligent Systems: Networks of Plausible Inference.* Morgan Kaufmann, San Mateo, CA (1988).

96. Perez, R.A., L.O. Hall, et al. "Evaluation of Machine Learning Tools Using Real Manufacturing Data." *International Journal of Expert Systems: Research and Applications*, To appear.

97. Pineda, F. J. "Dynamics and architecture in neural computation." *J. Complexity.* Vol. 4 (1988), pp. 216-245.

98. Pineda, F. J. "Generalization of back propagation to recurrent neural networks." *Phys. Rev. Lett.*, Vol. 59 (1988), pp. 2229-2232; also "Generalization of back propagation to recurrent and high-order networks." D. Z. Anderson, Ed. *Neural Information Processing Systems.* New York: American Institute of Physics (1987), pp. 602-611.

99. Politakis, P. and S.M. Weiss. "Using Empirical Analysis to Refine Expert System Knowledge Bases." *Artificial Intelligence.* Vol. 22 (1984), pp. 23-48.

100. Pople Jr., H.E. "Knowledge-based Expert Systems: The Buy or Build Decision." W. Reitman, ed. *Artificial Intelligence Applications for Business.* Ablex, Norwood, NJ (1984).

101. Quinlan, J.R. *C4.5: Programs for Machine Learning.* Morgan Kaufmann, San Mateo, CA (1992).

102. Reddy, D.R., L.D. Erman, R.D. Fennel, and R.B. Neely. "The HEARSAY speech understanding system: An Example of the Recognition Processs." Proc. IJCAI 3 (1973), pp. 185-193.

103. Rich, E. and K. Knight. *Artificial Intelligence*, 2nd edition. McGraw-Hill, NY (1991).

104. Roth, M. W. "Neural-network technology and its applications." *Heuristics*, Vol. 2, No. 1 (1989), pp. 46-62.

105. Roysam, B. and M. I. Miller. "Mapping deterministic rules to stochastic representations via Gibbs' distributions on massively parallel analog networks: Application to global optimization." D. Touretzky, G. Hinton and T. Sejnowski, Eds. *Proc. 1988 Connectionist Models Summer School.* San Mateo, CA: Morgan Kaufmann (1989), pp. 229-238.

106. Rumelhart, D. and D. Zipser. "Feature discovery by competitive learning." *Cognitive Science.* Vol. 9 (1985).

107. Sakaguchi, S., I. Sakai, and T. Haga. *Application of Fuzzy Logic to Shift Scheduling Method for Automatic Transmission.* FUZZ-IEEE, San Francisco, CA. (1993), pp. 52-58.

108. Sekiguchi, M. and K. Asalcawa. "Mobile robot control by a structured hierarchical neural network." *IEEE Control Systems Magazine*. Vol. 10 (1990), pp. 69-76.

109. Shafer, G. *A Mathematical Theory of Evidence*. Princeton, Univ. Press, Princeton, NJ (1976).

110. Shafer, G. and J. Pearl. *Readings in Uncertain Reasoning*. Morgan Kaufmann, San Mateo, CA (1990).

111. Shastri, L. "A connectionist approach to knowledge representation and limited inference." *Cognitive Science*, 12 (1988), pp. 331-392.

112. Shavlik, J. and J. Diettrich. *Readings in Machine Learning*. Morgan Kaufmann, San Mateo, CA (1991).

113. Shortliffe, E.H. *Computer-Based Medical Consultation: MYCIN*. Elsevier North-Holland, NY (1976).

114. Sun, G. Z., H. H. Chen, and Y. C. Lee. "Learning stereopsis with neural networks." M. Caudill and C. Butler, Eds. *Proc. IEEE First Int. Conf. Neural Networks*. Vol. IV, Piscataway, NJ: IEEE (1987), pp. 345-355.

115. Tanaka, K. and M. Sugeno. "Stability Analysis and Design of Fuzzy Control Systems." *Fuzzy Sets and Systems*, 45, No. 2 (1992), pp. 135-156.

116. Tank, D. W. and J. J. Hopfield. "Simple neural optimization networks: an A/D converter, signal decision circuit, and a linear programming circuit." *IEEE Trans. Circuits and Systems*. Vol. CAS-33 (1986), pp. 533-541.

117. Touretzky, D. and G. Hinton. "A distributed connectionist production system." *Cognitive Science*. Vol. 12 (1988), pp. 423-466.

118. Touretzky, D. and G. Hinton. "Symbols among the neurons: Details of a connectionist inference architecture." *Proc. Ninth International Joint Conference of Artificial Intelligence*. Los Angeles, CA, (Aug. 18-23, 1985), pp. 238-243.

119. Valverde, L. and E. Trillas. "On Modus Ponens in Fuzzy Logic." *The Fifteenth International Symposium on Multiple-Valued Logic,* Kingston, Ontario, Canada (1985).

120. van Melle, W. "A Domain-independent Production-rule System for Consultation Programs." *Proceedings IJCAI-79* (1979), pp. 923-925.

121. Waterman, D.A. "A Guide to Expert Systems." Addison-Wesley, Reading, MA (1986).

122. Weiss, S.M., C.A. Kulikowski, S. Amarel, and A. Safis. "A Model-Based Method of Computer-Aided Medical Decision-Making." *Artificial Intelligence*. 11 (1978), pp. 145-172.

123. Weiss, S.M. and C.A. Kulikowski. "Representation of Expert Knowledge for Consultation: The CASNET and EXPERT projects." P. Szolovits, ed. *Artificial Intelligence in Medicine*. AAAS Symp. Series, Westview Press, Boulder, CO. (1982), pp. 21-55.

124. Wenstop, F. "Applications of Linguistic Variables in the Analysis of Organizations," Ph.D. Thesis, U.C. Berkeley, CA (1975).

125. Whalen, T. and B. Schott. "Goal-Directed Approximate Reasoning in a Fuzzy Production System." Gupta et al., eds. *Approximate Reasoning in Expert Systems*. North-Holland, NY (1985).

126. Winston, P. H. *Artificial Intelligence*. 3rd edition, Addison-Wesley, Reading, MA (1992).

127. Yager, R.R. "Robot Planning with Fuzzy Sets." *Robotica* 1 (1983), pp. 41-50.

128. Yang, Q. and V.K. Bhargava. "Building expert systems by a modified perceptron network with rule-transfer algorithms." *Proc. International Joint Conference on Neural Networks*. San Diego, CA, (June 17-21, 1990), Vol. II, pp. 77-82.

129. Zadeh, L.A. "The Role of Fuzzy Logic in the Management of Uncertainty in Expert Systems." *Fuzzy Sets and Systems*. 11 (1983), pp. 199-227.

130. Zadeh, L.A. "Commonsense Knowledge Representation based on Fuzzy Logic." *Computer*. 16, (10) (1983), pp. 61-65.

131. Zadeh, L.A. "Review of Books: A Mathematical Theory of Evidence." *The AI Magazine*. Vol. 5, No. 3 (1984), pp. 81-83.

132. Zadeh, L.A. "A Simple View of the Dempster-Shafer Theory of Evidence and its Implication for the Rule of Combination." *The AI Magazine*. Vol. 7, No. 2 (1986), pp. 85-90.

133. Zadeh, L.A. "Outline of a Theory of Usuality Based on Fuzzy Logic, Berkeley Cognitive Science Report Series." U.C. Berkeley, CA (1986).

134. Zismanm, M. "Use of production systems for modeling asynchronous, concurrent processes." D.A. Waterman and F. Hayers-Roth Eds. *Pattern-Directed Inference Systems*. New York: Academic Press (1987).

2

APPLICATIONS OF FUZZY EXPERT SYSTEMS IN INTEGRATED OIL EXPLORATION*

F. Aminzadeh, Unocal Corporation
Anaheim, California, USA

Expert systems, though not widely used in the oil industry, have been the subject of a large volume of research and development activities in the industry and academia. Two reasons for limited practical usage of expert systems in oil exploration are (1) exploration in general is highly multi-disciplinary, and (2) the rules governing the exploration process are, for the most part, subjective. The combination of these two factors has made development of expert systems for solving practical exploration problems difficult. Recent advances in some areas of expert systems, coupled with the availability of cost effective and fast workstations, offer opportunities to overcome the two major obstacles. Specifically, using fuzzy logic networks in expert systems makes integration of different knowledge sources, implementation of inexact and qualitative rules (information), and self learning more practical.

---

* This chapter is adapted from the first chapter of *Expert Systems in Oil Exploration*, F. Aminzadeh and M. Simaan, eds, Society of Exploration Geophysicists, Tulsa, 1991.

2.1 INTRODUCTION

Application of expert systems in exploration dates back to the pioneering work of Hart et al. (1978) for mineral exploration and evaluation of mineral resources. Their system, PROSPECTOR, used an expert "consultant" to evaluate a body of geological information and compare the information against a set of rules and models developed for ore deposits and morphology of igneous rocks.

Since PROSPECTOR was introduced, numerous articles have been published describing expert system application in the oil industry. These articles reported on various systems that were in different stages of completion, from a mere concept, to a collection of rules, to a prototype, and rarely, to an actual working system in a commercial sense. Even rarer is a working system that has gained wide acceptance and is used extensively. For an extensive list of expert systems related work in the area of minerals, oil and gas exploration, and production over the last 10 years, i.e., see Aminzadeh & Simaan (1989), Simaan & Aminzadeh (1991), and Aminzadeh (1991).

Here, we are going to address two questions: Why expert systems are not widely used (in spite of a considerable number of publications) in the oil industry? Among possible answers I offer two major reasons: (1) the multidisciplinary nature of the problem and (2) the nonprecise (subjective) nature of the rules.

Exploration, in general, is a multidisciplinary type of activity. An expert system that claims to be capable of solving a given exploration problem would imply that all the rules, know-how, and information that are used in solving that exploration problem are incorporated in the system. However, developing an expert system that might contribute to the solution of a practical exploration problem, would be next to impossible because the expertise of several hundred people would have to be incorporated. Moreover, the computer requirement for such a system would be immense.

The other factor in oil exploration applications is that not only are we dealing with a multi-disciplinary type of activity, we are also involved in an industry in which most of the rules are subjective. Since our rules often are not exact and are undeterministic, to derive a mathematical representation of the relevant rules is difficult. For example, to establish a mathematical representation of the set of rules to distinguish between bright spots associated with hydrocarbon accumulation and hard streaks would be difficult. Furthermore, many rules would be required. Thus, using conventional mathematics, even to tackle a small subset of a large exploration problem would require hundreds or thousands of rules with immense computational requirements.

Fuzzy expert systems eliminate the need to introduce a large number of rules to represent a simple concept. Thus, exact rules (rules that do not involve elements of ignorance or fuzziness), if they ever exist, can be viewed as a limiting case of approximate reasoning. Using fuzzy rules inference becomes, as Zadeh (1989) says, a "process of propagation of elastic constraints." The end result of this process is a set of outcomes (conclusions) with certain degrees of elasticities. Introduction of Fuzzy-Lisp for knowledge representation has started a new trend for the development of expert systems.

Several attempts have been made to use fuzzy logic or evidential reasoning, which can be considered as a special case of fuzzy logic, in oil exploration. Griffiths (1984) used a fuzzy-set based pattern recognition approach for strata recognition from drilling response. Anand et al. (1988) report on their work in progress in which they use Dempster Shafer's "mixed initiative" approach to build an expert system with inexact reasoning (MIDST) that combines evidences from different origins. This system, designed to help geologists characterize hydrocarbon, plays both through determining the salient characteristics of the area and through comparing the area with analogous known plays [also see Pai (1988) on this expert system]. Aminzadeh et al. (1989) discuss integration of knowledge from different sources using evidential reasoning. Tabesh (1990) uses both fuzzy logic and evidential reasoning to find an optimum location for drilling by integrating data from different sources (also, see Popoli & Mendel (1989) on intelligent seismic signal processing, An & Moon (1990) on integrated exploration, Fang et al. (1991) on thin section mineral identification, and Lashgari (1991) on classification of geophysical data using fuzzy logic). Several other applications of fuzzy logic in expert systems for exploration are given below.

2.2 SEGMENTATION OF SEISMIC SECTIONS BASED ON TEXTURE USING FUZZY RULES

The object of Figure 2.1 is to divide the seismic section into Regions 1 through 4. The criterion for the division is the texture or character of the seismic signal. Based on differences in the texture, this segmentation might lead to a geologically meaningful interpretation. The following rules are used:

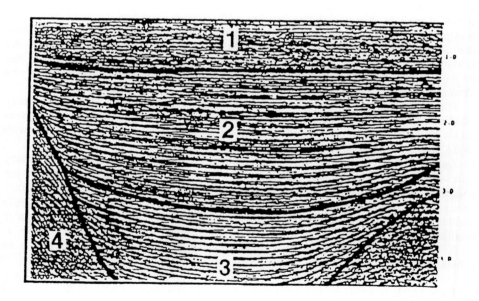

Figure 2.1. A manual segmentation of a seismic section into four regions of common signal character and texture.

Rule 1. The Depth Prejudice Rule: It is *unlikely* that a texel below 2 s should be classified as Region 1. It is *unlikely* that a texel above 1 s should be classified as Region 3. (Note that this rule is related to the effect of sediment compaction on seismic signal character.)

Rule 2. The Sequence Rule: In a vertical column of texels, if a texel directly above the texel in question is classified as Region 2, then it is *unlikely* that the texel in question belongs to Region 1. The same applies for Region 2 with respect to Region 3.

Rule 3. Region 4 Shape Rule: Region 4 is a solid figure instead of a layer. That is, it is composed of salt domes and shale ridges. It is *improbable* that a texel is in Region 4 if the texels above and below that texel are classified other than Region 4. Rule 3 is related to the expected topology of the image. That is, a diapir is expected to have nearly vertical sides.

Rule 4. The Neighbors Rule: If, among the nearest eight neighbors around a texel in question, the *majority (>4)* are assigned to a particular region, then it is *very likely* that the texel in question belongs to that region.

If analyzed carefully, note that these rules are unnecessarily restrictive. Rules are stated in this manner not because of precise requirements but, merely, because of the type of mathematics (conventional versus fuzzy) used to formulate the rules. What degree of accuracy is necessary as far as the two s rule is concerned? How about 2.01 s? As for the number of regions, is requiring four regions really important or even correct? Should all the samples be forced to belong to one and only one of those regions? In most cases, these rules can and should be modified. But more important, certain elements of these rules are qualitative. Thus, they are impossible to formulate with predicate calculus. For example, how can the rules involving expressions such as: *improbable, majority*, and *very likely* be formulated and implemented in the computer using conventional approaches?

These concerns are overcome by using fuzzy logic. The main aspect of fuzzy set theory is that an element can belong to different classes simultaneously with different degrees (grades of membership), $\mu(A)$. (Thus, a "fiscal conservative" can be a Democrat with $\mu = 0.8$, and a Republican with $\mu = 0.15$ and an independent with $\mu = 0.05$!).

Zadeh (1989) gives an excellent introduction to knowledge representation using fuzzy logic. He states "It is the qualitative mode of reasoning that plays a key role in the application of fuzzy logic." Using this type of logic, use of fuzzy predicates, fuzzy quantifiers, and fuzzy probabilities such as *unlikely, improbable, majority*, and *very likely* would be practical. Also, using fuzzy logic, the constraints could be elastic and thus more practical.

A fuzzy representation of **Rule 1** of Seismic Segmentation based on texture is as follows:

Rule 1f. It is *unlikely* that a texel below *approximately* 2 s should be classified as Region 1. It is *unlikely* that a texel above *approximately* 1 s should be classified as Region 3.

Among the implications of this rule are: (1) membership or classification in different regions is not clear cut, and a texel can belong to Regions 1 and 2 simultaneously with different degrees of membership, (2) the transition from above 2 s to below 2 s need not be abrupt, and (3) qualitative descriptors such as unlikely and approximately are acceptable.

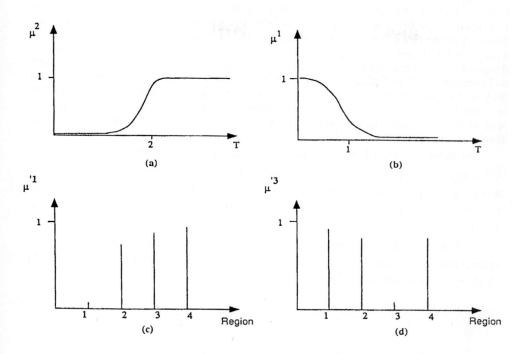

Figure 2.2. Membership Grade Distributions of Fuzzy Representations of Rule 1:
(a) below approximately 2 seconds,   (b) above approximately 1 second,
(c) unlikely to belong to Region 1,   (d) unlikely to belong to Region 3.

A more formal (but simplified) representation of Rule 1f is:

Rule 1f. If t texel = T with μT, then the texel belongs to Region i with μi.

Figures 2.2a and 2.2b show a possible distribution for $\mu 2$ and $\mu 1$ (below approximately 2 s and above approximately 1 s), respectively. Figures 2.2c and 2.2d show the distributions of $\mu 1$ and $\mu 3$ (unlikely to belong to Region i). For a more detailed treatment of the fuzzified rules (1 through 4 above), see Zhang & Simaan (1991).

2.3 INTEGRATING DATA FROM DIFFERENT DISCIPLINES USING EVIDENTIAL REASONING

Evidential Reasoning can be considered a special case of fuzzy logic. It is based on the Dempster-Shafer theory (DST) presented in Dempster (1967) and Shafer (1976), is an effective method to represent "ignorance," incomplete information, or inexact rules in expert systems. DST also provides a mechanism to handle conflicting data and rules (conflict management). For this reason and many others, an evidential reasoning approach is an ideal tool to integrate knowledge from different sources. For example, Expert 1 may believe that the chance of finding oil in a given prospect is at least 40 percent (indicating a range of 60 percent ignorance). Expert 2, on the other hand, may be of the opinion that the given prospect may contain oil or water with a chance of at least 60 percent. Using DST and combining the two pieces of evidence, the conclusion reached is that the chance of finding "oil" is between 40 and 64 percent, and the chance of finding "oil or water" is between 36 and 60 percent. The ignorance range is reduced from the original 60 and 40 percent to 24 percent. In the "Where are we going?" section a more detailed example illustrates the calculations.

The main difference between DST and the conventional probability theory is that in DST knowledge of probability of A (or belief in A), P(A), or Bel(A) does not provide the knowledge of probability of "not A" or Bel($\overline{A}$). In DST, Bel($\overline{A}$) <1-Bel(A) and 1-P(A)-P($\overline{A}$) is defined as the ignorance range I(A). By definition (Bel(A) + I(A)) is referred to as the plausibility of A, Pl(A). Thus, Bel (A), which is also referred to as the credibility of A, and Pl(A) form the lower and upper probability boundaries for the occurrence of A. Figure 3 illustrates probability functions (P = 0.3) of the classical theory, (P = 0.2 to 0.5) of Evidential Reasoning and (membership grade function, μ_A) of fuzzy logic.

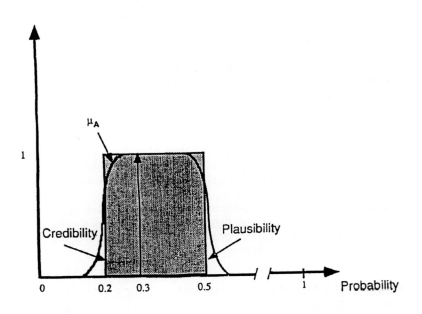

Figure 2.3. Comparison of probability functions in Classical Theory (P=0.3), Evidential Reasoning (P-0.2, 0.5) and Fuzzy Logic (μ_A = Membership Grade Function).

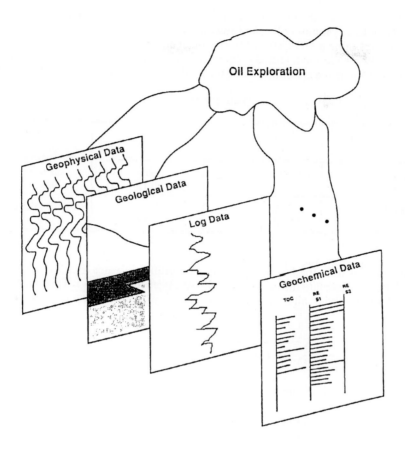

Figure 2.4. Integration of geophysical, geochemical, and geological data.

Here another hypothetical exploration problem is discussed to further illustrate the way the evidential reasoning approach can bring information from different sources (frames) to reduce "ignorance" and come up with tighter bounds for the "solution space." The example also shows how "conflicts" and contradictory information are handled. Suppose three different disciplines: geophysics, geology, and geochemistry, are used to predict the type of saturation in a sandstone, schematically illustrated by Figure 2.4. The following are the initial conclusions: Based on the geophysical data, we have a 30 percent chance for gas saturation and a 40 percent chance for "water or oil" saturation. Geological data indicates a 40 percent chance for "gas or oil." Geochemical data give a 50 percent chance for "water" saturation and a 30 percent chance for "oil" saturation. Using the evidential reasoning jargon, these probabilities are the lower bounds or the credibility factors. There are 30, 60, and 20 percent ranges of "ignorance" for geophysical, geological, and geochemical data involved.

Table 2-1 shows how the knowledge from geological and geochemical "frames" are combined. Note that combining the first row and column results in conflict (with a 0.2 factor). Also, combining a credibility of 0.3 for oil and 0.4 for gas or oil gives a credibility of 0.12 for oil. When the unidentified column is combined with the "gas or oil" row, the result is "gas or oil" with the credibility being the product of those of the row and the column.

TABLE 2-1

Geochem / Geology	Water (0.5)	Oil 0.3	Unidentified (0.2)
gas or oil (0.4)	conflict 0.20	oil 0.12	gas or oil 0.08
unidentified (0.6)	water 0.30	oil 0.18	unidentified 0.12

The next step is to normalize the resulting credibilities: water (0.3), oil (0.3), and gas or oil (0.08) to eliminate the conflicting one (0.2) called Cred $_{Conf}$. This normalization is done by multiplying all the numbers by

$$K = \frac{1}{1 - Cred._{conf}} = 1.25$$

Thus, the geological and geophysical data combined give the following odds:

gas or oil	0.1
water	0.375
oil	0.375
unidentified	0.15

From these odds we give the following credibility and plausibility factors: gas (0.0, 0.25), water (0.375, 0.525) and oil (0.375, 0.525). Note that the ignorance range of the combined data is .15, being smaller than the ignorance range of both geological data (0.6) and geochemical data (0.2). Also note how the odds with conflicting outcomes are handled. Table 2-2 shows the next step of combining geochemical and geological data with the geophysical data. Final results after necessary calculations are performed, both before and after normalization, are given in Table 2-3.

Table 2-3 gives the following lower and upper bounds for each type of saturation: gas (0.097, 0.194), water (0.339, 0.474), and oil (0.39, 0.564). Observe that all the ignorance factors or boundaries of the solutions (the difference between the upper and lower bounds) are further tightened as all three types of data are combined. The results of integration are the same irrespective of the order in which different knowledge sources are combined. This is demonstrated in Aminzadeh (1991).

TABLE 2-2

Geophysical / Geochem + Geology	Gas 0.3	Water or Oil 0.4	Unidentified 0.3
gas or oil 0.1	gas 0.03	oil 0.04	gas or oil 0.03
water 0.375	conflict 0.1125	water 0.15	water 0.1125
oil 0.375	conflict 0.1125	oil 0.15	oil 0.1125
unidentified 0.15	gas 0.045	water or oil 0.06	unidentified 0.045

TABLE 2-3

SATURATION	BEFORE NORMALIZATION	AFTER NORMALIZATION
Gas	0.075	0.097
Gas or Oil	0.03	0.039
Water	0.2625	0.339
Water or Oil	0.06	0.077
Oil	0.3025	0.39
Unidentified	0.045	0.058
Conflict	0.225	--

2.4 FUZZY RULES IN STRATIGRAPHIC INTERPRETATION

Earlier we demonstrated some of the inexact and subjective nature of many of the rules used in a particular expert system. To further illustrate the point, an example from sequence stratigraphy shows that many of the rules and definitions used in oil exploration, involve descriptors that are difficult to approximate. One such approximation to the rules in stratigraphic interpretation is presented by Bettazzoli et al. (1990).

Three types of deltaic facies, prodelta, delta-front, and alluvial are characterized according to their seismic response by the following guidelines adopted from the classic paper by Brown and Fisher (1977). The italics words highlight the inexact and fuzzy nature of these rules:

Prodelta and distal delta-front, barrier facies—reflection patterns for these facies in dip sections are *horizontal to steeply inclined*, oblique, layered patterns within a zone that *ranges from poorly layered to reflection-free* or *locally chaotic*. Oblique reflections *may* converge (and baselap) downward (basinward). In strike sections, the facies *commonly* exhibit convex-upward, *conformable drape-to-mounded-chaotic*, or reflection-free patterns with *some evidence* of channel or gully erosion. On the relict shelf, the prodelta reflections are *discontinuous except for a few strong reflections*, amplitudes are *generally low* except for reflections with *moderate continuity*. and spacing is *very erratic*.

Delta-front, barrier-bar facies—reflection patterns in dip sections are *horizontal to slightly inclined*, parallel-layered *near* the base, grading upward *irregularly* into *chaotic or reflection-free* patterns with *common convex-upward diffractions* and *poorly defined*, mounded reflections. *Subtle, inclined reflections* within chaotic zones *may* represent delta-front or barrier-bar offlap and, hence, *may* constitute internal time lines. In a strike section, the basal reflections of the zone *exhibit drape patterns* and *local chaotic*, to *reflection-free*, zones *display* subtle, *parallel-layered to draped*, reflections and *abundant* diffractions. Basal reflections exhibit *strong* continuity, but *continuity diminishes upward* in the unit. *The best continuity* occurs in dip sections. Amplitudes are *moderate to high* in basal, high-continuity reflections, but low in chaotic intervals; spacing is *moderately uniform* in basal reflectors, but erratic in the upper part of the zone.

Alluvial, delta-plain facies—reflection patterns in dip sections are *principally horizontal*, parallel, *rarely divergent layered to locally reflection-free*; locally, erosional channels *may be* inferred. In strike sections, the reflections are *weak, parallel-layered to subtle-mounded, chaotic-to-drape* patterns. Continuity of reflections range from *excellent to fair* in dip sections, but continuity is *poor to fair in strike sections*; amplitude is variable (high in continuous reflections and poor in chaotic zones); and spacing is *very regular* in zones of *high-continuity reflections* but *irregular* in the remainder of the unit.

Given the nature and structure of rules such as these, implementing them in an expert system based on classical (exact) mathematics is nearly impossible. The obvious solution is a system based on fuzzy inference. The following data (facts) are assumed available:

> A1. dip section reflection pattern,
> A2. strike section reflection pattern,
> A3. dip section reflection continuity,
> A4. strike reflection continuity,
> A5. dip section reflection strength, and
> A6. strike section reflection strength.

Table 2-4 (opposite page) outlines the described rules.

Subsets of X_1, X_2, X_3, with all the possible elements in them are defined. For example:

> X_1: (Horizontal, oblique, vertical, layered, convex upward, convex downward, drape mounded).

Note that A_1 and A_2 are fuzzy quantities defined over the fuzzy subset X_1 with membership grades of f_1^i and f_2^i, $i = 1, 2, \ldots I$, where I is the number of elements in X_1. The subsets of X_2 and X_3 may be defined as:

> X_2: (locally chaotic, discontinuous reflections, continuous reflections)

TABLE 2-4

	Prodelta & Distal Data Front Barrier Facies	Delta Front, Barrier Facies	Alluvial-Delta-Plain Facies
A_1 Dip section reflect pattern	Horizontal to steeply inclined, Oblique (layered to poorly layered pattern)	Horizontal to slightly inclined, parallel layered near the base, common convex upward reflection	Principally horizontal, parallel
A_2 Strike section reflection pattern	Convex upward	Drape pattern for the basal reflections	Parallel layered to suble-mounded, chaotic to drape
A_3 Dip section reflection continuity	Locally chaotic	Better continuity than the strike section	Excellent to fair
A_4 Strike section reflection continuity	Mostly discontinuous reflection	Basal reflections exhibit strong continuity diminishing upward in the unit	Poor to fair
A_5 Dip section ref. strength	low reflection to reflection free	Poorly defined mounded reflection, suble inclined reflection within the chaotic zone.	High
A_6 Strike section reflection strength	Weak reflection to reflection free, low amplitude in relict shelf except for those with moderate continuity	High to moderate amplitude in basal, high continuity reflection but low in chaotic areas	Weak

and

X_3: (low reflection amplitude, high reflection amplitude, reflection free).

Given the fuzzy information and rules, the input data goes through fuzzy inference mechanism, the result of which is the fuzzy classification of the data into possible stratigraphic types prodelta, delta front, alluvial (Y, Y_2, and Y_3) with different membership grades (g_1, g_2, g_3, ...). Figure 2.5 shows a possible fuzzy inference network for this example.

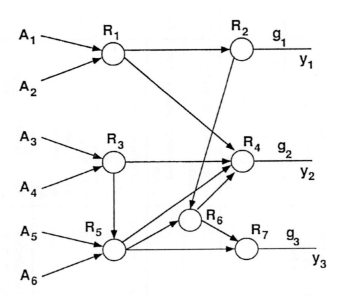

Figure 2.5. Fuzzy inference network.

Figure 2.5 is only for illustration and is not meant to incorporate the rules of Table 2-4. However, the figure does conceptually show how a series of fuzzy rules can be combined to reach a fuzzy outcome describing the membership grade of the input data to different stratigraphic types of Y_1, Y_2, Y_3.

This example illustrates the basic concepts of knowledge representation using fuzzy logic without going into any detailed theoretical discussion. For a rigorous treatment of this subject, see Eshera and Barash (1989) and Chen, et al. (1990). In these references rule-based inference and knowledge representation is accomplished by fuzzy mesh-connected systolic arrays and fuzzy petri nets, respectively. Alternatively, neural networks have been used for lithologic classification, Guo, et al. (1992) and Lorenzetti (1992), for geophysical interpretation, Calderon-Macias and Sei (1993).

2.5 CONCLUSION

To make expert systems more practical and more useful, the power of some of the unconventional techniques should be unleashed. If our rules are fuzzy and/or if we take advantage of evidential reasoning, fewer rules are needed to explain a certain situation, and data and knowledge from different sources can be integrated more effectively. Also, subjective concepts can be formulated and implemented in our systems. Bringing these tools to our domain of expertise enables development of expert systems capable of solving meaningful exploration problems.

REFERENCES

1. Aminzadeh, F., F. S. Wong, and E. H. Ruspini. "A practical view of expert systems for oil exploration: Integration of multiple knowledge sources." In M. Simaan and F. Aminzadeh, Eds., *Advances in geophysical data processing.* Vol. 3, *Artificial intelligence and expert system in petroleum exploration.* JAI Press (1989), pp. 1-17.

2. Aminzadeh, F. "Where Are We and Where Are We Going." In F. Aminzadeh and M. Simaan, Eds., *Expert Systems in Exploration.* Society of Exploration Geophysicists (1991), pp. 3-32.

3. Aminzadeh, F. and M. Simaan. *Expert Systems in Exploration,* Society of Exploration Geophysics (1991).

4. An, P. and W. M. Moon. "Application of fuzzy set theory for integrating geological and geophysical data." *60th Internat. Mtg. Soc. Expl. Geophys. Expanded Abstracts* (1990), pp. 366-369.

5. Anand, T., G. Biswas, M. Pai, C. Kendal, R. Canon, P. Morgan, and J. Bezdek. "XX Hydrocarbon exploration using a knowledge based approach." Presented at the *Ann. AAPG-SEPM-EMD-DPA Convention, AAPG Bull.* **72.** No. 2 (1988), p. 155.

6. Bettazzoli, P., M. T. Galli, and P. Rochhini. " EXSSTRA: Integrated knowledge-based system for stratigraphic interpretation." *60th Internat. Mtg. Soc. Expl. Geophys., Expanded Abstracts* (1990), pp. 328-331.

7. Brown, L. F. Jr. and W. L. Fisher. "Seismic-stratigraphic interpretation of depositional systems, Examples from Brazilian Rift and Pull-Apart Basins" In E. E. Payton, Ed., *Seismic stratigraphic applications to hydrocarbon exploration: AAPG's Memoir 26* (1977), pp. 218-248.

8. Chen, S. M., J. S. Ke, and J. F. Chang. "Knowledge representation using fuzzy petri nets." *Inst. Elec. Electron. Eng. Trans. Knowledge and Data Eng.*, **2.** No. 3 (1990), pp. 311-319.

9. Dempster, A. P. "Upper and lower probabilities induced by a multivalued mapping, *Annals Mathematical Statics,* **38.** No. 2 (1967), pp. 325-329.

10. Eshera, M. A. and S. C. Barash. "Parallel rule—based fuzzy inference on mesh-connected systolic arrays." *Inst. Elec. Electron. Eng., Expert*, **4**. No. 4 (1989), pp. 27-35.

11. Fang, J. H., H. C. Chen, and D. Wright. "A Fuzzy Expert system for Thin-Section Mineral Identification." In Aminzadeh, F. & Simaan, M., Ed., *Expert Systems in Exploration*, SEG (1991), pp. 179-202.

12. Griffiths, C. M. "An example of the use of fuzzy-set based pattern recognition approach to the problem of strata recognition from drilling response." Presented at the *27th International Geological Congress*, Moscow, 1984, and reprinted in *Pattern Recoognition & Image Processing*, Aminzadeh, F., Geophysical Press (1989), pp. 504-538.

13. Guo, Y., R. O. Hansen, and N. Harthill. *Feature Recognition from Potential Fields Using Neural Networks*. 62nd International Manufacturing Society Expl. Geophysicists Extended Anbstracts (1992), pp. 1-5.

14. Hart, P. E., R. O. Duda, M. T. Enaudi. "PROSPECTOR - A computer based consultation system for mineral exploration, *Mathematical Geology*, **10** (1978), pp. 589-610.

15. Lashgari, B., "Fuzzy Classification with Application to Geophysical Data." In F. Aminzadeh and M. Simaan, Eds., *Expert Systems in Exploration*, SEG (1991), pp. 161-178.

16. Lorenzetti, E. A. *Predicting Lithology from Vp and Vs using Neural Networks*. 62nd International Manufacturing Society Expl. Geophysicists Extended Abstract (1992), pp. 14-17.

17. Calderon-Macias, C. and M.K. Sei. *Geophysical Interpretation by Artificial Neural Systems*. 63rd International Manufacturing Sciety Expl. Geophysicists Extended Abstract (1993), pp. 254-257.

18. Pai, M. L. M. "A knowledge-based system approach for hydrocarbon prospect analysis." *Ph.D. Thesis, University of South Carolina* (1988).

19. Popoli, R. F. and J. M. Mendel. "Heuristically constrained estimation for intelligent signal processing." In M. Simaan and F. Aminzadeh, Eds, *Advances in geophysical data processing*, Vol. 3, *Artificial intelligence and expert system in petroleum exploration*. JAI Press (1989), pp. 107-133.

20. Shafer, G. "A mathematical theory of evidence." *Princeton University Press* (1976).

21. Simaan, M. and F. Aminzadeh, Eds. *Advances in geophysical data processing*, Vol. 3, *Artificial intelligence and expert system in petroleum exploration*. JAI Press (1989).

22. Simaan, M., Z. Zhang, P. L. Love. "Image processing and knowledge based methods for segmentation of a seismic section based on a signal character." In F. Aminzadeh, Ed., *Pattern recognition and image processing*. Geophysical. Press (1987), pp. 389-425.

23. Tabesh, E. "DRILL-PAT: An Expert System to design an optimum drilling pattern." *Ph.D. Dissertation*, Syracuse University (1990).

24. Zadeh, L. A. "Knowledge representation in fuzzy logic." *Inst. Elec. Electron. Eng. Trans. Knowledge and Data Engineering*, **1**. No. 1 (1989), pp. 89-100.

25. Zhang, Z. and M. Simaan. "Control system for SEISIS, A rule-based system for segmentation of a seismic section based on texture." In *Simaan, M. and Aminzadeh, F., Eds., Advances in geophysical data processing*, Vol. 3, *Artificial Intelligence and Expert System in Petroleum Exploration*, JAI Press (1989), pp. 135-173.

3

HARDWARE APPLICATIONS OF FUZZY LOGIC CONTROL

M. Jamshidi, R. Marchbanks[1], E. Kristjansson[2], K. Kumbla, R. Kelsey[3], and D. Barak
University of New Mexico
Albuquerque, New Mexico, USA

Fuzzy logic and fuzzy expert control systems are currently among the most active research and development areas of artificial intelligence. Thanks to tremendous technical advances and many industrial applications developed by the Japanese, fuzzy logic enjoys an unprecedented popularity. The object of this chapter is to describe a number of hardware implementations of fuzzy control systems. The chapter is organized in five sections: Section 3.1 provides a brief introduction to fuzzy logic and fuzzy control. Section 3.2 describes a real-time fuzzy control and tracker for a laser beam system. Section 3.3 discusses a fuzzy controller for a power system generation unit. A fuzzy control application for a robotic manipulator is given in Section 3.4, and Section 3.5 illustrates the experimental

---

[1]Currently with Los Alamos National Laboratory, Los Alamos, NM, USA
[2]Currently with Polaroid Corporation, Cambridge, MA, USA.
[3]Currently with Department of Computer Science, New Mexico State University, Las Cruces, NM, USA.

experiences for real-time fuzzy control of a model train on a circular path. A fuzzy PI controller for a thermoelectric cell used for non-chlorofluorocarbon (CFC) air conditioning and refrigeration systems is given in Section 3.6. Finally, Section 3.7 presents and discusses some conclusions.

3.1 INTRODUCTION TO FUZZY LOGIC AND FUZZY CONTROL

Most of the current applications of fuzzy logic are fuzzy expert control systems. Fuzzy controllers are expert control systems that smoothly interpolate between otherwise crisp (or predicate logic-based) rules. Rules fire to continuous degrees and the multiple resultant actions are combined into an interpolated result. The processing of uncertain information and energy savings through the use of common-sense rules and natural language statements provide the basis for fuzzy logic control (or "fuzzy control," for short). The use of sensor data in practical control systems involves several tasks that are usually performed by a human in the decision loop, e.g., an astronaut adjusting the position of a satellite or putting it in the proper orbit, a driver adjusting an air conditioning unit, etc. The performance of all such tasks must be based on evaluation of pertinent data according to a set of rules that the human expert has learned from experience or training. Often, if not most of the time, these rules are not crisp (based on binary logic), i.e., they involve common sense and human judgment in the decision-making process. Such problems (or "judgment calls") can be addressed by a set of fuzzy variables and rules which, if calculated and executed properly, can make expert decisions.

In an attempt to translate the crisp knowledge of a process to a fuzzy knowledge, i.e., to go through the process of *fuzzification*, one needs to represent the input and output variables of a plant by a number of linguistic label members of some fuzzy set, e.g., the set of bright images on the focal plane of a telescope, or a set of small voltages, etc. Fuzzy sets may be represented by a mathematical formulation often called the *membership function*. This function gives a degree or grade of membership within the set. The membership function of a fuzzy set A, denoted by $\mu A(X)$, maps the elements of the universe X into a numerical value within the range [0,1]:

$$\mu A(X)$$
$$X \longrightarrow [0,1]$$

Sometimes a membership function is also called a *possibility* function but not a *probability* function. To distinguish between these two functions, one should note that probability functions represent the frequency of an event, while the possibility of an event is a grade of membership (or similarity) attached to a single attribute for the same event. In control system applications, membership values are actually measures of degree of causality in an input-output mapping. Fuzzy logic gave a new definition to the causality in dynamic systems. Within this framework, a membership value of zero corresponds to a value which is definitely *not* an element of the fuzzy set, while a value of 1 corresponds to the case where the element definitely *is* a member of the set. In fuzzy logic, like binary logic, operations such as union, intersection, complement, OR, AND, etc. are all defined (see Jamshidi et al., 1993,[1] Chapters 2 and 3).

The basic structure of a fuzzy controller takes the form of a set of IF-THEN rules whose *antecedents* (IF parts) and *consequents* (THEN parts) are membership functions. Consequents from different rules are numerically combined (typically union via MAX) and are then collapsed (typically taking the centroid or center of gravity of the combined distribution) to yield a single real number (binary) output. Within the framework of a fuzzy expert system, as with regular expert systems, typical rules can be the result of a human operator's knowledge, e.g.:

"If the *Temperature* is Hot, and Pressure is High then increase the *Current* to a Medium level."

In this rule, Hot , High and Medium are fuzzy (linguistic) variables. Such natural language rules can then be translated into typical computer language type statements such as:

IF (A is A1 and B is B1 and C is C1 and D is D1) THEN (E is E1 and F is F1)

Using a set of rules such as these, an entire finite number of rules can be derived in the form of natural language statements, as if a human operator were performing the controlling task. In any practical system, such as an air-conditioning unit, the user or operator often fine-tunes, tweaks, or adjusts the knobs until the desired cool (or hot) air can be felt with the desired speed. Such operator knowledge can be utilized in the design of a fuzzy controller for an air-conditioning unit system. One of the most common ways of designing fuzzy controller is through "fuzzy rule-based systems." The controller shows the processes of *fuzzification*, i.e., binary-to-fuzzy transformation and *defuzzification*, i.e., fuzzy-to-binary transformation.

There is an alternative method of implementing a fuzzy control regime which is similar to a conventional adaptive control law. A standard 1-, 2- or 3-term controller such as P, PI, PD, or PID can be used and then, through some adaptation loop, the conventional controller's gains K_p and K_d (for example, in a PD control law, $u = K_p y + K_d (dy/dt)$), can be adjusted through fuzzy rules.

A fuzzy set, as first proposed by Zadeh,[2] corresponds to a set whose boundaries move. For example, the set of all *"young men"* corresponds to a variety of men at many different ages, say from 18 to 35, whose difference is in the membership value (degree of similarity or belonging). In fuzzy sets, a continuum of possible choices can be used to describe an imprecise term. The approach advocated by Zadeh is based on the premise that the human thinking process is vague rather than exact. The notion addressed here is based on the human's ability to extract information from masses of inexact data. The human experience represents an important source of such approximate thinking. Thus the theory of fuzzy sets helps one to translate a linguistic model of human thinking into a fuzzy algorithm.

Fuzzy logic corresponds to the logic of approximate reasoning. Like classical (crisp) logic, it can handle symbolic manipulation, but more importantly, it handles numerical data as well. Figure 3.1 shows a block diagram for a fuzzy control system. As shown, the analog plant's output is recognized by sensory elements and the corresponding voltage from the sensor is converted into digital words by an analog-to-digital (A/D) converter, often installed on the digital computer control board. Once the system's output is digitized, it can be changed into a finite number of fuzzy (linguistic) variables through the FUZZIFIER by assigning fuzzy membership functions to various ranges of the output's universe of discourse. Once the digital (crisp) values have been fuzzified, they can be used in the premises (IF parts) of the fuzzy inference engine within the fuzzy expert system, as shown.

The resulting fuzzy decision (control) variables in linguistic form will be converted back to crisp values through the process of defuzzification (DE-FUZZ in Figure 3.1). The crisp control signal, still in digital form, is finally converted back into analog signal to be used to actuate the analog plant. Currently, four possible real-time control architectures can be implemented for real-time self-tuning fuzzy control systems. Figure 3.2 represents all four implementations. As seen, one is based on Togai InfraLogic Fuzzy-C software,[3] one is based on Bell Helicopter's FULDEK,[4] one is based on the NeuraLogix FMC board,[5] and the fourth is based on Helikon's IDL-747 imbedded fuzzy-controller board. Note that the first three architectures are based on a PC, while the fourth one is an imbedded system. Other possible imbedded systems not shown here include Togai's SBFC (single-board fuzzy controller), Aptronix-FIDE software, and a Motorola chip to download fuzzy rules for an imbedded control card.

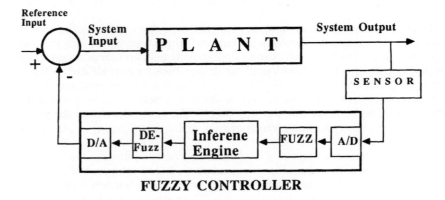

Figure 3.1. Block diagram for a typical fuzzy control system showing fuzzifier (FUZZ), defuzzifier (DE-FUZZ), and inference engine.

3.2 FUZZY CONTROL OF A LASER TRACKING SYSTEM

In this section a real-time fuzzy controller is designed in a laboratory setting to track a helium-neon laser beam. Here, fuzzy logic is used to control a two-axis mirror gimbal for aligning a laser beam onto the center of a quadrant detector. A PC-based controller was implemented with a simple rule set to control and stabilize the device. A comparison was made between a non-fuzzy proportional controller and the fuzzy controller. Experiments were performed to study the tracking characteristics and the effect of rule-pruning on the fuzzy controller response. Figure 3.3 shows the experimental set-up.

3.2.1 Problem Statement

Communication systems that use line-of-sight transmitting and receiving are desired. This is especially true from a military point of view with respect to submarine communi-

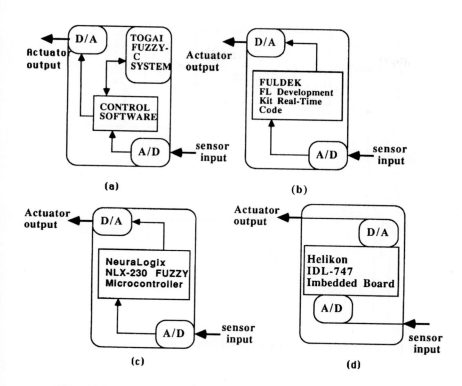

Figure 3.2. Four real-time architectures for fuzzy control systems: (a) Software implementation; (b) Bell Helicopter's FULDEK; (c) NeuraLogix's NLX-230 FMC; and (d) Helikon's IDL-747 card.

cation. A submarine requires the ability to communicate with other platforms but still remain "invisible." The requirement for such a system is robust tracking that will maintain the communication link between platforms, given a variety of disturbances such as movement, bumps, obstructions, and atmospheric or hydrospheric perturbations. This section presents a demonstration of fuzzy control applied to such a problem. It is an attempt to attain a proof of principle for application of fuzzy control to optical systems.

3.2.2 Design Approach

A two-axis mirror gimbal is set up to steer a laser beam onto a quadrant detector so that the beam lies on the center of the detector. Electronics sense any error in the position of the beam relative to the center of the detector in both the x and y directions. These errors are converted to a digital format for use by the fuzzy controller, which is PC-based. The control software processes the error information using fuzzy logic and decides on the appropriate response for the motor driver. The driver supplies the required power to run the motors and reposition the beam (see Figure 3.3).

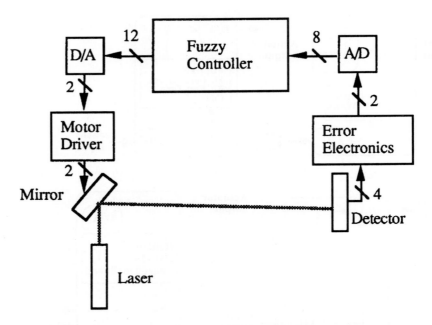

Figure 3.3. Block diagram of the laser beam closed-loop control system.

3.2.3 System Hardware

For feedback information, a quadrant detector was used to sense the error between the actual position of the beam and the desired position. The beam center should be on the center of the detector. A quadrant detector is simply four photosensitive semiconductor elements that produce a voltage proportional to the amount of photon flux incident on each one of them. A centered beam will have one quarter of its light on each quadrant and the potentials will be equal. As the beam moves off-center, the proportions change. Error in the horizontal and vertical directions can be separated by simple electronics that reduce the single two-dimensional problem into two one-dimensional problems. Based on the size and price of commercially available quadrant detectors, a decision was made to build one using solar cells. This constructed detector had a 16-square-cm area and a cost of less than $10.

The mirror consisted of a movable gimbal with horizontal and vertical degrees of freedom. Mounted on the mirror were two DC motormikes to control the elevation and azimuth of the mirror. One complication in using DC motors is that they have a nonlinear speed as a function of armature voltage. The main problem lies in the *dead band* region, where the motor requires a minimum voltage to begin moving. Any voltage below this turn-on voltage will not drive the motor. This nonlinear characteristic is shown, along with the required control response, in Figure 3.4. A driver employing this response will always keep the armature voltage above the turn-on voltage when there is non-zero error.

This response however, adds instability and can cause limit cycling near zero voltage. That is, due to the very high slope of the control response curve near the origin, the motor can switch between a forward and reverse direction, causing oscillation. Part of the objective in the design was to reduce or eliminate limit cycling with the fuzzy logic controller.

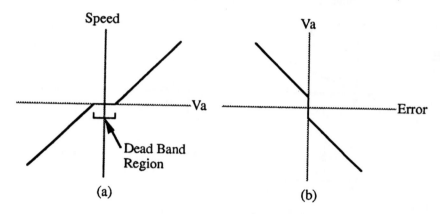

Figure 3.4. (a) Motor characteristics; (b) Desired armature voltage response.

The controller was implemented using a PC with a 386 microprocessor. A Keithley-Metrabyte 8-bit analog-to-digital converter was used to put the analog error in the digital format required by the computer. The input of the A/D converter has a range of –5 volts to 5 volts with a resolution of 8 bits over this range. The error signal was conditioned with amplifiers to fill this range and acquire the full 8-bit resolution. After processing, the resulting digital response signal was converted to an analog voltage by a Keithley-Metrabyte 12-bit digital-to-analog (D/A) converter.

3.2.4 System Software

Designing the required software to perform fuzzy inference can be a very involving task, but fortunately there are a few packages that have been written do this. One of these packages, Togai InfraLogic Fuzzy-C, was used to write the primary code to perform the fuzzy arithmetic. This software is programmed in a language developed by the manufacturer which allows the user to easily construct membership functions and a rule base. The resulting code is compiled into C by a special compiler. This software performs all the required fuzzy operations, including defuzzification, and produces a crisp output. A driver program, written in C to communicate with the A/D and D/A converters, also acquires data from the experiments. Listings of the Fuzzy-C code, the A/D and D/A converter addresses, and driver program are listed in appendices A, B, and C, respectively of a former report (see Reference 1, Chapter 7).

To represent the error input to the controller, a set of linguistic variables was chosen to represent 5 degrees of error, 3 degrees of change in error, and 5 degrees of armature voltage. Membership functions were constructed to represent the input and output values' degrees of truth for each of these linguistic variables. These membership functions are shown in Figure 3.5.

52   Soft Computing

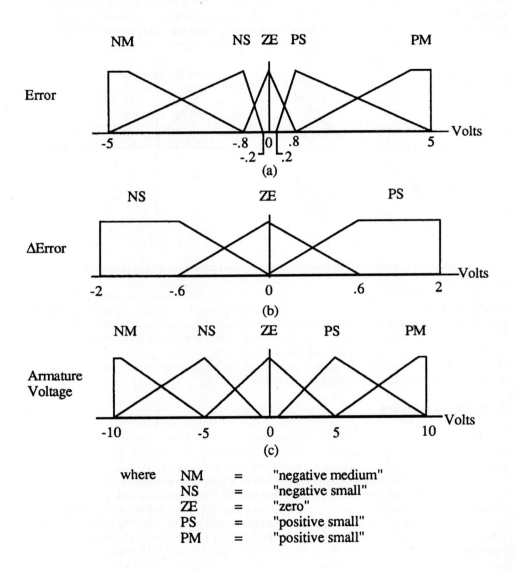

Figure 3.5 Membership functions: (a) Input error; (b) Input change in error; (c) Output armature voltage.

Hardware Applications of Fuzzy Logic Control

The input error membership functions contains a very narrow zero error (ZE) region. A faster convergence was expected from this since, for a small error, the degree of "medium" will be large and the armature voltage will correspondingly be larger than if ZE were broad. Also, the "medium" membership functions do not intersect at zero for both the error and armature voltage. This was done to help suppress overshoot and limit cycling.

Two sets of rules were chosen. These "Fuzzy Associative Memories," or FAMs, are a shorthand matrix notation for listing the rule set. A linguistic armature voltage rule is fired for each pair of linguistic-error and linguistic-change-in-error variables. For example, if the error is PM and the change in error is ZE, then the armature voltage is NM. Because overlap exists between the fuzzy variables, more than one rule can fire simultaneously.

The FAMs are shown in Figure 3.6. Figure 3.6(a) shows the full set of 15 rules. Figure 3.6(b) shows the set after pruning. The number of fuzzy IF-THEN rules can be potentially very large. The number of rules can be determined by the number of input variables i in the antecedents and the number of output variables o in the consequents. The number of rules R can then be defined as

$$R = N1 \cdot N2 \; N3 \ldots Ni \cdot o$$

where Ni represents the number of membership functions (linguistic labels) assigned to the ith input variable. For the laser beam system, we have $R = 5 \times 3 \times 1 = 15$ rules. Large number of rules can be reduced by several methods. One is through aggregation of rules (see Reference 1, Chapter 5); the other is by the fusion of input variables coming from the sensors, i.e., reduction of the antecedent variables; and a third method is to keep track of the rules being fired during real-time inferencing of the fuzzy control and to eliminate those rules which are either not used at all or used very little. In this project, experimental testing allowed the elimination of eight of the 15 rules, as shown in Figure 3.6(b). In addition, two rules were modified to compensate for the loss of these eight. The two new rules were:

1. IF Error is NS and Derriere is NS THEN Armature Voltage is PS

2. IF Error is PS and Derriere is PS THEN Armature Voltage is NS

(a)

		Error				
		NM	NS	ZE	PS	PM
ΔError	NS	PM	PM	PS	ZE	NS
	ZE	PM	PS	ZE	NS	NM
	PS	PS	ZE	NS	NM	NM

(b)

		Error				
		NM	NS	ZE	PS	PM
ΔError	NS		PS		ZE	
	ZE	PM		ZE		NM
	PS		ZE		NS	

Figure 3.6. Fuzzy Associative Memories (FAMs):
(a) Full set (15 rules); (b) Pruned set (7 rules).

3.2.5 Experimental Results

Two experiments were performed to test the integrity of the controller, as illustrated in Figure 3.7. First, the center-locating ability was examined by placing the beam on the detector with the center of the beam offset in the positive x direction from the center of the detector. The system was started with data taken at 100 Hz for 4 seconds. The rate of convergence to the detector center was observed as well as the overshoot and oscillation. This experiment was performed for a non-fuzzy proportional feedback controller, for the fuzzy controller with 15 rules, and for the pruned fuzzy controller with 7 rules.

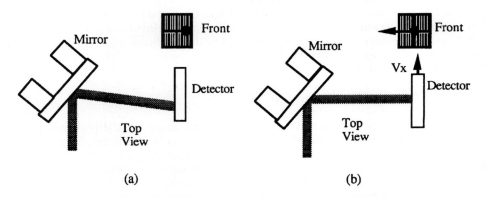

Figure 3.7 Experimental arrangement: a) Center locating; b) Tracking.

The second experiment tested the system's tracking ability. The detector was mounted on a linear positioner with the beam centered on the detector. The positioning stage was driven at 0.2 mm/s in a direction normal to the beam incidence. The distance from the mirror to the detector was 1.5 m. Again, data was taken at 100 Hz for 4 seconds while this experiment was performed. A non-fuzzy proportional feedback controller, the fuzzy controller, and the pruned fuzzy controller were tested using this scheme.

The implementation involved six experiments to test the system. For center-locating ability, a non-fuzzy controller, a fuzzy controller and a pruned fuzzy controller were used. These three controllers were again used in the tracking experiment. Results are shown in Figure 3.8.

<u>Center locating</u> With the beam center initially offset in the positive x direction, the proportional feedback controller, the fuzzy controller, and the pruned fuzzy controller were all used to drive the system. The results for 4 seconds of time are shown in Figures 3.8(a)-(c). When comparing the fuzzy controllers to the non-fuzzy controller, the y-error is omitted to prevent the graphs from becoming too busy.

Figure 3.8(a) shows the behavior of the non-fuzzy controller. Both x-error and y-error are shown for 4 seconds of time. The error in the x direction is maximized to 5 volts until the beam begins to overlap the center. It quickly converges, overshoots and oscillates as the controller reverses directions. The y-error remains fairly stable until the beam reaches the center. The oscillation in this error is due to the misalignment of the 4 quadrants. When the

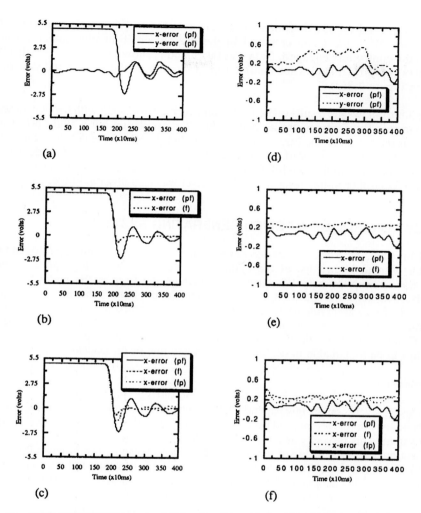

Figure 3.8. Experiment results: (a) Non-fuzzy (proportional feedback) x and y-error; (b) Non-fuzzy x-error vs. fuzzy x-error; (c) With pruned rules; (d) Tracking non-fuzzy x and y-error; (e) Tracking non-fuzzy x-error vs. fuzzy x-error; (f) Tracking with pruned rules.

beam crosses over to the remaining quadrants, the abrupt change due to misalignment causes the oscillation in error.

With the fuzzy controller engaged, the overshoot is dramatically reduced and oscillation is effectively eliminated. This plot is shown in Figure 3.8(b). Next, the pruned fuzzy set of rules was implemented and tested. The graph in Figure 3.8(e) shows more overshoot and oscillation than that of the unpruned controller, but the results are better than those obtained with the non-fuzzy controller.

Tracking The second set of experiments was performed to test the tracking ability of the controllers. With the beam centered in the desired position, the detector was moved and the controllers engaged. Data was taken and is shown in Figures 3.8(d)-(f) The non-fuzzy controller shows substantial oscillation in both the x-error and y-error, as shown in Figure 3.8(d)

Figure 3.8(e) shows the results of the fuzzy controller x-error along with the non-fuzzy controller x-error. The fuzzy controller does reduce the oscillation but a slight offset is observed. This indicates that the fuzzy controller lags slightly behind the detector. Finally, the pruned fuzzy controller is tested. As in the center-locating test, the pruned set does not reduce the oscillation as well as the full fuzzy set did. However, it does perform better than the non-fuzzy controller in this respect. A slight offset was also observed here.

3.3 FUZZY CONTROL OF A POWER GENERATING SYSTEM

A real-time fuzzy logic controller was designed for a synchronous generator driven by a DC-motor without knowledge of the systems transfer function. The purpose of the controller was to keep the output frequency and voltage steady at 60 Hz and 120 V_{rms}, respectively, with only the knowledge of the system input/output behavior. From this information the fuzzy logic rule base was formed and implemented into software with the Togai InfraLogic Shell. The controller was tested in real time for different loads on the generator. Comparisons with adaptive crisp controllers, underway at this point, have shown fuzzy control to be remarkable in dealing with load and line disturbances.

3.3.1 Introduction

In the last few years only a few papers[6-10] have been published on the use of fuzzy logic in power systems. For this reason, and because of the strong nonlinearity of the components in power systems, this project was an interesting challenge to the authors.

This section describes the use of fuzzy logic to control an electric power generator and its prime mover in real-time. The system is stand-alone and is therefore not dependent on the behavior of any other generator. Since a stiff power network is not present in the system, any load change will have a sizable effect on both the output frequency and the output voltage.

The purpose of the fuzzy logic controller is to keep the frequency at 60 Hz and the voltage at 120 V_{rms} for any changes in load, and to remove all setpoint deviations as fast as possible.

3.3.2 System Set-Up

The system consists of the following units:

1. DC-motor, which is the prime mover;
2. Synchronous machine; and
3. Fuzzy logic controller (computer, control circuits, and measurement circuits).

Figure 3.9 shows a block diagram of the system set-up.

Hardware Applications of Fuzzy Logic Control

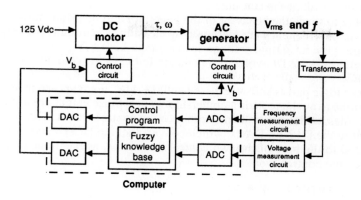

Figure 3.9. Block diagram of system set-up.

The DC-motor is connected in compound fashion, which provides a simple means of speed control by changes in the rheostat resistant. Because of the need to control the speed with a computer, the rheostat is replaced by a power transistor, which functions as a current controller. The base voltage of the transistor is supplied by the DAC (see Figure 3.10) in the computer.

The generator is driven by the DC-motor and supplies an output of three-phase 208 V_{rms} at 60 Hz. The change in voltage output is controlled by an excitation field, provided by an external DC source and, to some extent, by the speed of the rotor. The control of the excitation field is implemented with a transistor in a fashion similar to the DC-motor (Figure 3.10). The transistor controls the magnitude of the current

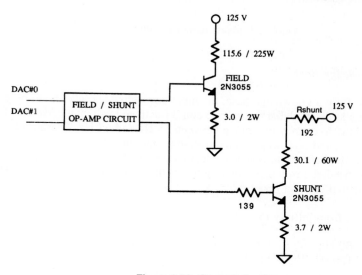

Figure 3.10. Control circuitry.

passing through the field windings and, as before, the base voltage of the transistor is supplied by the DAC in the computer.

To be able to sense the behavior of the generator and its prime mover, the output voltage and output frequency are measured by measurement circuits (see Figure 3.11). These circuits give an output voltage in the range of 0-5 V_{DC}, proportional to the measured values. These DC-voltage signals are fed in to the computer through a pair of 8-bit ADCs.

The fuzzy logic part is software-based and the ADC readings are sent to the fuzzy logic subroutine to evaluate their membership values; those values are used to execute appropriate rules, which consequently produce the control values. The control values are then sent to the 8/12-bit DACs, which convert them to DC-voltage signals. These signals are fed to the base of the two control transistors, as mentioned above, which in fact closes the control loop.

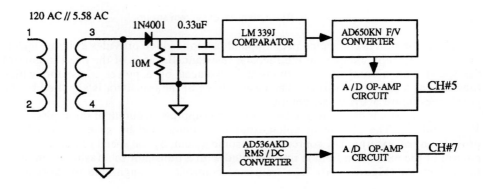

Figure 3.11 Measurement circuitry.

3.3.3 Fuzzy Knowledge Base

To be able to form a knowledge base, the input and output signals had to be given a working range. It was decided that the voltage could fluctuate in a range between 110 V_{rms} and 130 V_{rms}, and the frequency in a range between 55 Hz and 65 Hz. These ranges were then divided into sets, and each set given membership functions. The next step was to form a knowledge base. But before that was possible, measurement of the input/output behavior of the system was needed. From these measurements the fuzzy rules were created. Here is an example of the making of a fuzzy rule:

Changing load from infinity to 1000W:

Frequency: 60 Hz -> 57 Hz => $\Delta F = 3$ Hz
Voltage: 120 VAC -> 113 VAC => $\Delta V_{AC} = 7 V_{AC}$

Control actions to get frequency back to 60 Hz and voltage to 120 VAC:

I_{shunt}: 0.310 VDC -> 0.261 VDC => ΔI_{shunt} = -0.049 ADC
V_f : 54.35 VDC -> 56.8 VDC => ΔV_f = 2.45 VDC

Then the following rule is generated from the fuzzy sets:

IF ΔF is L **AND** ΔVAC is M

THEN DIshunt is N **AND** ΔVf is P

where L stands for the membership function Low, M for Medium, N for Negative, and P for Positive.

In this way, the entire set of rules was formed, based on the change in input and output. The final rule base is as follows. (L=Low, M=Medium, H=High, N=Negative, Z=Zero, P=Positive)

Rule 1: **IF** ΔF is L **AND** ΔVAC is L
THEN ΔIshunt is N **AND** ΔVf is N

Rule 2: **IF** ΔF is L **AND** ΔVAC is M
THEN ΔIshunt is N **AND** ΔVf is Z

Rule 3: **IF** ΔF is L **AND** ΔVAC is H
THEN ΔIshunt is N **AND** ΔVf is P

Rule 4: **IF** ΔF is M **AND** ΔVAC is L
THEN ΔIshunt is Z **AND** ΔVf is N

Rule 5: **IF** ΔF is M **AND** ΔVAC is M
THEN ΔIshunt is Z **AND** ΔVf is Z

Rule 6: **IF** ΔF is M **AND** ΔVAC is H
THEN ΔIshunt is Z **AND** ΔVf is P

Rule 7: **IF** ΔF is H **AND** DVAC is L
THEN ΔIshunt is P **AND** ΔVf is N

Rule 8: **IF** ΔF is H **AND** ΔVAC is M
THEN ΔIshunt is P **AND** ΔVf is Z

Rule 9: **IF** ΔF is H **AND** ΔVAC is H
THEN ΔIshunt is P **AND** ΔVf is P

Membership functions are shown in Figures 3.12-3.14.

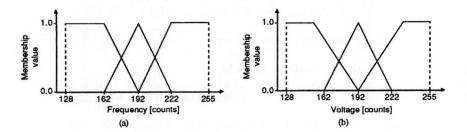

Figure 3.12. Membership functions for input values of controller.

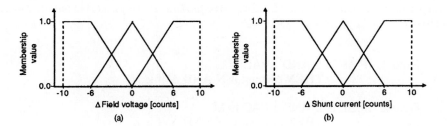

Figure 3.13. Membership functions for output values of controller using an 8-bit DAC.

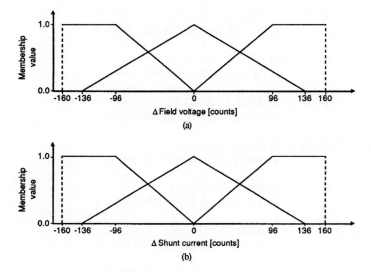

Figure 3.14. Membership functions for output values of controller using a 12-bit DAC.

The fuzzy sets, their membership functions, and the fuzzy rules were then assembled into a fuzzy knowledge base by using the TIL Shell in MS Windows.[3] The TIL Shell output file was compiled to generate a C-subroutine, which was then linked to the main control program.

3.3.4 Experiment

To test the system, a load unit was connected to the output of the generator. This load unit is a box of parallel connected resistors, with each resistor activated by a switch. Figure 3.15 shows the load unit diagram.

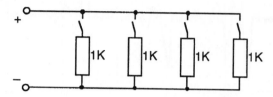

Figure 3.15. Load unit.

Note that each additional load does not add resistance on the output, but does decrease the resistance and in that way demands power. Therefore the generator must increase its power output, which means more power to the DC-motor. Results from different load changes are shown in the following section.

3.3.5 Results

The frequency and voltage system responses are shown in Figures 3.16 and 3.17, respectively. In Figure 3.16 (8-bit case), it can be seen that signals do settle around their optimal values but their fluctuation is too big to be satisfactory. The problem is in the

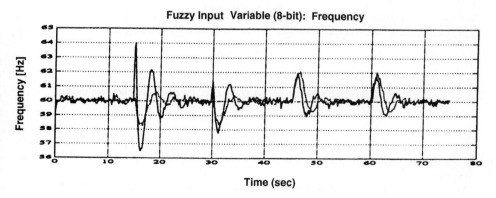

Figure 3.16. Frequency response for 8- and 12-bit DAC.

sensitivity of the control signals, shown in Figures 3.10 and 3.11 (8-bit case); i.e., even though the system outputs are still not completely settled, the control signals have already settled because change in the input is too small to achieve any change in the output. Therefore, the 8-bit DAC has insufficient precision for this particular circuit.

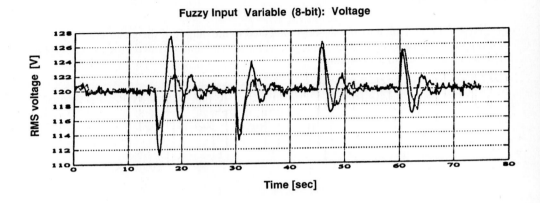

Figure 3.17. Voltage response for 8- and 12-bit DAC.

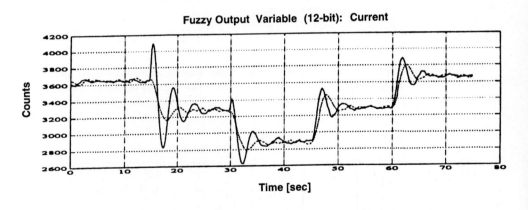

Figure 3.18. Control signal from 8- and 12-bit DAC for control of field voltage.

To solve this problem the number of DAC bits is increased to 12, which gives a much wider control resolution. Figures 3.8 and 3.9 (12-bit case) show the results of the 12 bit control signals.

As shown in Figure 3.16, the fluctuations around optimal values were minimized with the 12-bit DAC, which supports the use of a more expensive DAC.

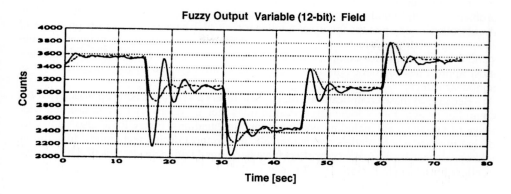

Figure 3.19. Control signal from 8- and 12-bit DAC for control of shunt current.

3.4 FUZZY CONTROL OF A ROBOT MANIPULATOR

A *robot* is a reprogrammable multifunctional mechanical manipulator designed to move materials, parts, tools, or special devices through planned trajectories for the performance of a variety of tasks. It is a computer-controlled manipulator consisting of several relatively rigid links connected in series by revolute, spherical, or translational joints. One of these links is typically attached to a supporting base while another link has an end that is free and equipped with a tool known as the end-effector for manipulating objects or performing assembly tasks. Mechanically, the robot is composed of an arm, a wrist subassembly, and a tool. It is designed to reach a workpiece located within some distance or workspace determined by the maximum and minimum elongations of the arm. The dynamic equations of the robot are a set of highly nonlinear, coupled differential equations containing a varying inertia term, a centrifugal and Coriolis term, a frictional term, and a gravity term. Movement of the end-effector in a particular trajectory at a particular velocity requires a complex set of torque functions to be applied by the joint actuators of the robot. The exact form of the required functions of actuator torque depend on the spatial and temporal attributes of the path taken by the end-effector, and on the mass properties of the links and payload, friction in the joints, etc. Figure 3.20 shows an educational Rhino robot under consideration.

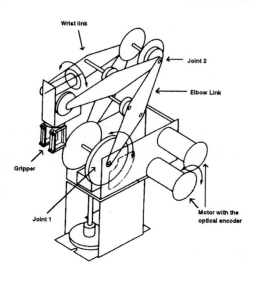

Figure 3.20 Schematic of a Rhino robot.

The nonlinear dynamics governing robot motion present a challenging control problem. A traditional linear controller cannot effectively control the motion of the robot. A controller based upon the theory of nonlinear control is better suited for the problems of robot manipulation. Unfortunately, nonlinear differential equations are plagued by substantial requirements for computation and have an incomplete theory of solution. Thus, most approaches to robot controller design have suffered due to the complications of nonlinear effects.

Because of these complications, fuzzy logic offers a very promising approach to robot controller design. Fuzzy logic offers design rules that are relatively easy to use in a wide range of applications, including nonlinear robotic equations. Fuzzy logic also allows for design in cases where models are incomplete, unlike most design techniques. In addition, microprocessor-based fuzzy controllers have performed with data streams of 8 bits or less to allow for a simple design.

3.4.2 Robot Models and Control

The dynamic equation of a robotic manipulator can be described by the nonlinear differential equation (1). For the purpose of simulation a two-link planar robotic manipulator is considered.[11]

$$M(\theta)\ddot{\theta} + C(\theta,\dot{\theta}) + G(\theta) = \tau \tag{3.1}$$

Where M is a two-dimensional matrix of inertia terms, C is a vector of centrifugal and Coriolis terms, G is a vector of gravity terms and is a vector of joint torques. θ, $\dot{\theta}$ and $\ddot{\theta}$ are the joint angular position, velocity and acceleration terms. The mathematical expression for a two-link robot is given by the following terms.[5]

Inertia matrix:

$$M(\theta) = \begin{bmatrix} a_1 + a_2 Cos\theta_2 & a_3 + 0.5a_2 Cos\theta_2 \\ a_3 + 0.5a_2 Cos\theta_2 & a_3 \end{bmatrix} \tag{3.2}$$

Coriolis and centrifugal torque vector:

$$C(\theta,\dot{\theta}) = \begin{bmatrix} (a_2 Sin\theta_2)(\dot{\theta}_1\dot{\theta}_2 + 0.5\dot{\theta}_2^2) \\ (a_2 Sin\theta_2)0.5\dot{\theta}_1^2 \end{bmatrix} \tag{3.3}$$

Gravity and loading vector:

$$G(\theta) = \begin{bmatrix} (a_4 Cos\theta_1) + a_5 Cos(\theta_1 + \theta_2) \\ a_5 Cos(\theta_1 + \theta_2) \end{bmatrix} \tag{3.4}$$

The parameters a_1, a_2, a_3, a_4, and a_5 account for the inertia and gravity that influence the nonlinear motion. Solutions to these nonlinear differential equations typically require several specialized techniques, such as Lyapunov theory, each addressing a part of the solution but none addressing the complete solution.

Fuzzy Logic Control Fuzzy logic provides a means of dealing with nonlinear functions. A fuzzy controller was designed to simulate the performance of the two-link Robotic manipulator model. The membership functions were developed for the effect of position errors and velocity parameters for the two links of the robot. A membership function for the output of the controller, i.e., the joint torque, was also defined. This was developed using Togai InfraLogic software.[3] The membership function of the fuzzy controller was first approximated by studying the response of a traditional PD controller, and then tuned to achieve the best response by trial and error. A set of 30 rules was developed, which is shown in Appendix A. Thus, the membership function incorporates the characteristic and dynamic of the manipulator.

Implementation The dynamics of the system are simulated using a software developed in C language. (The program is listed in Appendix 14-B of reference.[1] The main function first defines the various global and local variables, and initial and final time of the simulation. The des_traj() subroutine defines the desired trajectory (step input) and the controller() subroutine gives the output of the fuzzy controller. It then calls the subroutine robo_state(), which defines the dynamic of the two-link manipulator. The dynamics of the robot are solved by the subroutine rk4(), which is basically a fourth-order Runge Kutta solution method. The out_file() converts the output in a suitable form that can be used to generate a Matlab-compatible file by a print_matlab() subroutine. This forms the driver program. The .til file which defines the fuzzy controller in Togai InfraLogic software is first converted into C code and merged with the driver program, then compiled and linked to form the executable code. The output data file generated by the simulation is used to plot the trajectory using Matlab software.

Simulation Results Figure 3.21 shows a comparison of the fuzzy controller vs. the PD controller. A response to a step input is plotted against time. Figure 3.21(a) shows the fuzzy controller response and Figure 3.21(b) shows the PD controller response. Simulation shows that the fuzzy controller has a rise time of 0.3 seconds, whereas the PD controller rise time is 0.4 sec. The fuzzy controller is able to improve the response by reducing the overshoot and at the same time decreasing the rise time. Note that various sets of gains were tried in the case of the PD controller to obtain an optimal performance. It has been seen that if the gains of the PD controller are increased to improve the rise time, there is an inherent increase in the rise time. Also, one set of gains of the PD controller does not provide the same performance in all operating ranges.

Fuzzy logic attempts to bypass the difficulties that accompany the solutions to nonlinear differential equations. Fuzzy logic allows us to set up a controller that is not entirely based on a complete description of the robot.

3.4.3 Baseline Hardware Design

Figure 3.22 provides an overall block diagram of the Rhino robot and control hardware. The robot has five links and a gripper; the three links used for controlled movement include the shoulder, elbow, and wrist. Each link was moved with a DC

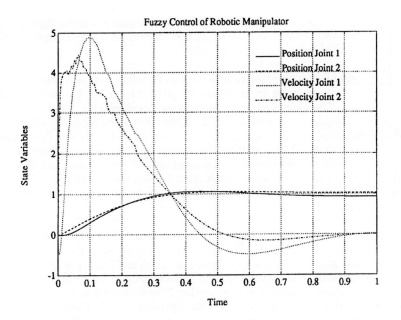
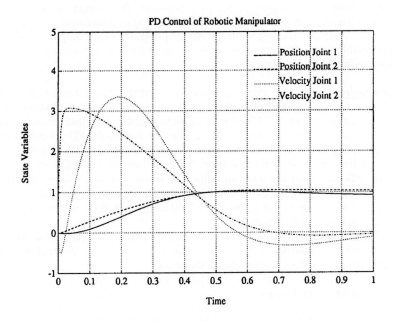

Figure 3.21. (a) Fuzzy controller simulation response of a robotic manipulator; **(b)** PD controller response of a robotic manipulator.

Hardware Applications of Fuzzy Logic Control

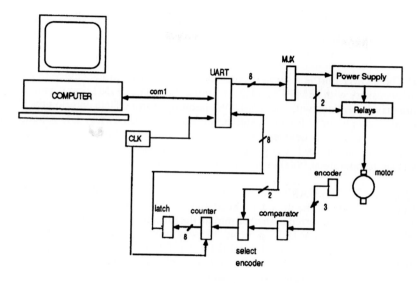

Figure 3.22. Schematic of the hardware.

servo motor rotating about an axis at each joint. The computer system used was a Packard Bell 386 PC. The remaining hardware consisted of elements for power control (driver), data acquisition, and communications.

Power Control The driver for the DC motors consisted of a Fluke 4265A programmable power supply, control circuits, and relays. Two 8-bit control words were used to control the driver. The control words were originally sent by the computer through the RS-232C COM1 port in serial format to the external universal asynchronous receiver/transmitter (UART). The UART then delivered the control words in parallel to the programmable power supply and MUX. The first 8-bit control was used to program the power-supply voltage level, thus allowing the supply to deliver voltages ranging from -16 to $+16$ volts in 0.25 volt increments. The second 8-bit control word controlled the time at which the power supply switched output settings and signaled the MUX to activate one of the relays. The relays were required to switch power to only one motor at a time. Thus, at any point in time, only one of the links could be activated.

Data Acquisition The data acquisition hardware provided the position feedback required by the software controller. Optical encoders were used to detect rotational movement of a motor and send six analog pulses for one complete revolution. Each pulse corresponded to 0.12 degrees of movement. The analog pulses then passed through a Schmitt-trigger comparator to give a clean pulse for digital circuit interface. The pulses incremented the counter and thus provided a measurement of total displacement. At every ten milliseconds, a crystal clock triggered the sampling of the counter. The count information was latched, loaded into the UART as parallel data, and then transmitted serially to the computer through the RS-232C COM1 port.

Communications The communications hardware provided the necessary elements for transporting data between the computer and external hardware. The primary hardware components used for communications included the COM1 serial port and the Universal Asynchronous Receiver Transmitter (UART).

COM1 port The COM1 port is a serial, asynchronous communications device that comes as a standard accessory with IBM PCs and most compatibles. The COM1 port was supported by a software driver provided with the MS-DOS operating system. Data entering the COM1 port is made available to the software hosted on the computer. Likewise, data generated by the software may be sent to devices external to the computer through the COM1 port interface. The data entering or leaving the port had to conform to the RS-232C standard. The port was set to operate at 9600 baud with eight data bits, two stop bits, and no parity.

UART Implementation The UART shown in Figure 3.23 is a 40-pin TR1602 by Western Digital Corporation. Four of the pins were available for RS-232C option settings. The NO PARITY pin (pin 35) was asserted high to eliminate the presence of the parity bit in the serial data. The NUMBER OF STOP BITS pin (pin 36) was asserted high to designate two stop bits for the serial data. The pins for NUMBER OF DATA BITS (pin 37 and pin 38) were both asserted high to set the number of data bits to eight. The other possible choices for NUMBER OF DATA BITS were five, six, and seven data bits. Eight more pins on the

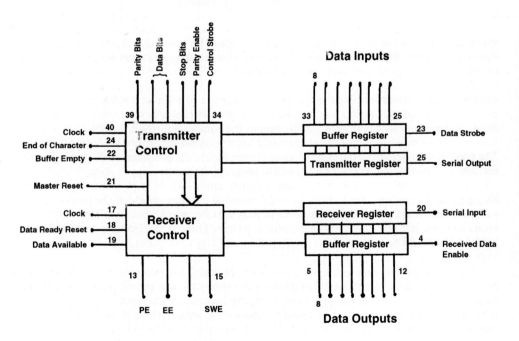

Figure 2.23. Functional Block Diagram of the UART.

UART were dedicated for TTL compatible, 8-bit parallel data interface.[12] The CLOCK pin (pin 40) of the UART was required to pulse at a frequency 16 times the baud-rate setting.

UART Transmitter Configuration The count information arrived at the UART as a parallel 8-bit data word awaiting to be transmitted serially to the COM1 port. When the DATA STROBE pin (pin 23) was asserted low, it signaled the UART to load the 8-bit data word into its transmitter buffer register. The TRANSMITTER BUFFER EMPTY pin (pin 22) provided a signal to the COM1 port indicating that the transmitter buffer register was empty (logic one). The END OF CHARACTER pin (pin 24) provided a signal to the COM1 port indicating when transmission was stopped.

During operation, the UART first transmitted TTL-compatible serial data through the SERIAL OUTPUT pin (pin 25) to the MC1488 IC. The MC1488 then converted the TTL signal to an RS-232C format compatible with the COM1 port.

UART Receiver Configuration Serial data originating from the COM1 port conformed to the RS-232C standard. The serial data then passed through the MC1489 where it was converted to a signal that was TTL compatible for interface with the UART. The TTL-compatible serial data then arrived at the SERIAL INPUT pin (pin 20) of the UART. The UART converted the serial data into parallel 8-bit data words capable of interfacing with the driver for the motors.

Bit-Rate Generator The UART required a stable clock signal to be applied to the CLOCK pin (pin 40) in order to transmit and receive serial data precisely at 9600 baud. Furthermore, the clock signal was required to operate at 16 times the frequency of the baud rate. The clock signal used was generated with a Motorola MC14411 bit-rate generator IC, which is essentially a frequency divider. A 5-volt source connected to the bit-rate generator enabled the clock to pulse at a rate of 1.832 MHz using a crystal oscillator. Different frequencies were available from the bit-rate generator IC for different parts of the system.

3.4.4 Software Implementation of the Fuzzy Controller

The controller software was divided into two primary modules written in the C language: the fuzzy control module and the communications module.

Fuzzy Control Module The first step in fuzzy controller design was the selection of input and output variables. The designated input variables were ERROR (robot link position error) and VELOCITY (velocity of the robot link). The output variable was chosen to be VOLTAGE (drive voltage output by the power supply). Each variable was accompanied with a set of membership functions, shown in Figure 3.24.

During actual operation, the computer read in the robot position data and then computed position error and velocity. The fuzzy controller fuzzified the input quantities through algorithms that operated on the input data as specified by the membership functions. Next, the fuzzified input quantities passed through a series of IF-THEN decision rules that formed the main body of the fuzzy controller. Thus the fuzzy controller routinely assessed the current state of the robot and determined which control action was most appropriate. Defuzzification was applied using the voltage output variable, and a control action was selected. The control action was then passed through the interface to the external power supply driver.

70   Soft Computing

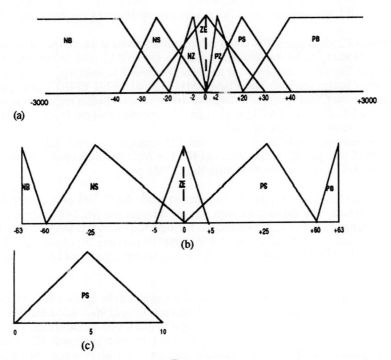

Figure 3.24. Fuzzy membership functions for: (a) Position error,
(b) Output voltage and (c) Velocity.

Table 3-1. Control words for the Driver.	
Function	**Code**
Shoulder Relay LSB 0000	MSB 1010
Elbow Relay LSB 0000	MSB 1011
Wrist Relay LSB 1000	MSB 1010
Positive Sign LSB 1000	MSB 1001
Negative Sign LSB 0000	MSB 1001

The output control signal from the computer is made up of two 8-bit control words that occupied two control cycles for each action completed. The first control word was used either to select one of the three relays or to change the voltage polarity output of the programmable power supply. This action was identified by bit 7, the most significant bit (MSB), and bit 6. When bit 7 was set to 1 and bit 6 set to zero, then either one of the relays was selected or the voltage polarity output of the power supply was set to the desired output. Bits 5, 4 and 3 were used to designate which relay to select or to set the output voltage polarity. When bit 7 was reset to zero, the control circuit then activated one of the motors of the robot. Bit 7 was used to strobe the programmable power supply and thus activate the programmed voltage output. The codes for the different actions are shown in Table 3-1.

The second control word was identified with both bit 7 and bit 6 set to 1. The desired voltage level output of the programmable power supply was selected by bits 0 (zero) through 5. Bit 7 was then set to zero in the next 8-bit control word to strobe the power supply and thus activate the newly selected power supply output voltage. Correct power supply operation required the use of twos-complement arithmetic with inverted logic.

The design of the fuzzy controller algorithm was accomplished with the Togai InfraLogic Fuzzy-C Compiler.[3] Input to the Togai Fuzzy-C Compiler consisted of a fuzzy source code with membership functions and a set of IF-THEN decision rules. (See Appendix 14-C of Jamshidi et al., 1993.[1] The Fuzzy-C Compiler converted the fuzzy source code into standard C source code which then passed through a C compiler to produce an executable code for the fuzzy controller. Trial experiments of the executable code were conducted to calibrate the membership functions until robot trajectory overshoot was suppressed and rise time was kept to a minimum.

__Communications Module__ A communications routine sampled the count information in the receive register every ten milliseconds. The fuzzy controller operated on this data and then output the results to a transmit register which was used to drive the DC motors. The program is listed in Appendix 14-D of Jamshidi et al., 1993.[1]

3.4.5 Testing and Results

Two tests were conducted to compare controller performance on robot arm trajectory. In both tests, the gain of the PD controller was set to provide the best performance in all ranges of operation. In the first test, each controller was set to traverse a trajectory of 200 counts in both positive (clockwise) and negative (counterclockwise) directions (Note: One count is equal to 0.12 degrees of rotation of the link). The test results for the trajectory of the shoulder, elbow, and wrist are shown in Figure 3.25 The results show that the fuzzy controller is able to move the robot arm smoothly to the desired position without overshoot. The PD controller, however, generated overshoot in the trajectory. It is interesting to note that when the robot link approached its final destination, its velocity remained high (steep trajectory curve) with PD control, but began to decrease much earlier with fuzzy control.

In the second test, the controllers were compared under various ranges of operation. The PD controller with suitable gain improved trajectory response by reducing the overshoot to less than four counts. However, the PD controller was unable to move the motor when the links were initially only one or two counts away from the final destination. The primary cause for this effect was the nonlinear friction in the robot joints. This resistance could be overcome with an increase in controller gain, but then the existing trajectory overshoot would be intensified.

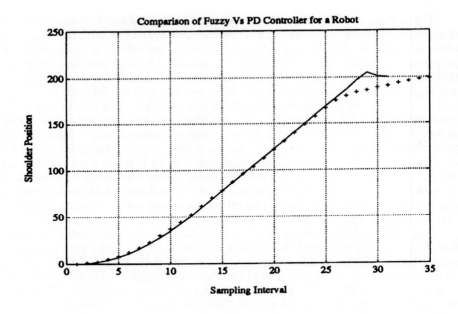

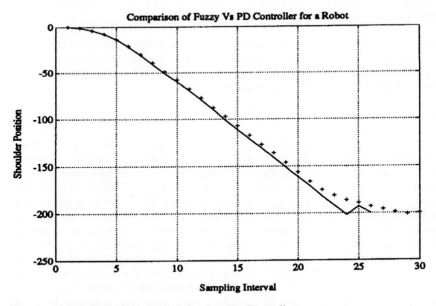

'+++' : Fuzzy Controller.   '----' : PD Controller.

Figure 3.25a. Experimental results for the fuzzy control of the robot manipulator—shoulder motion of the robot.

Hardware Applications of Fuzzy Logic Control

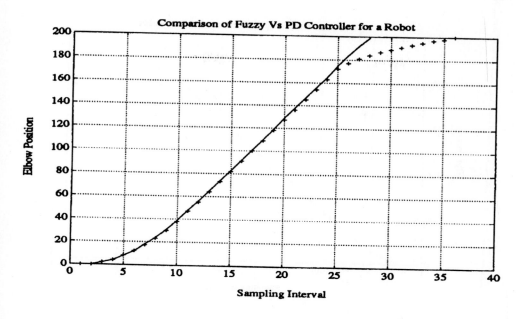

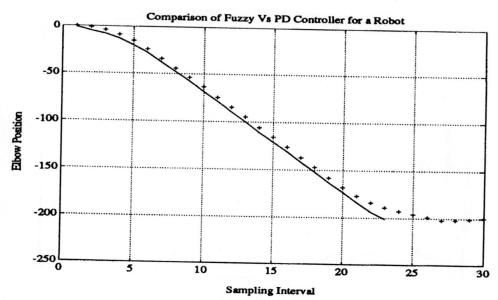

'+++' : Fuzzy Controller.   '----' : PD Controller.

Figure 3.25b. Experimental results for the fuzzy control of the robot manipulator—elbow motion of the robot.

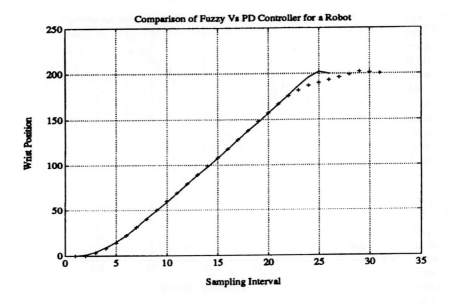

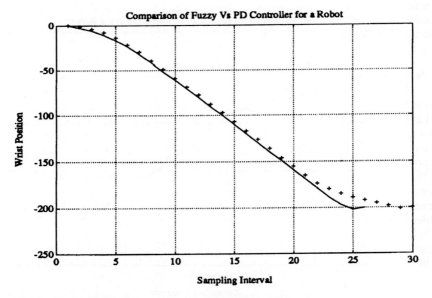

'+++' : Fuzzy Controller.   '----' : PD Controller.

Figure 3.25c. Experimental results for the fuzzy control of the robot manipulator—wrist motion of the robot.

The fuzzy controller was able to overcome the problems encountered by traditional PD control and perform better in all conditions tested. Results showed that the maximum overshoot with fuzzy control was held to a count of one. The fuzzy controller was also able to actuate the motor and travel a short distance of one count, unlike the PD controller.

The fuzzy controller exhibited robustness in performance against non-ideal effects, such as robot inertia, Coriolis effect, and gravity. These effects influence the velocity, position, and acceleration of the robot links, and thus negatively impact controller performance. However, with fuzzy control, only one set of fuzzy membership functions was required to guard against these non-ideal effects and operate all links of the robot. This result is not shared by the conventional PD controller, which required different gains to accommodate for the differences between links.

3.5 FUZZY ACCELERATION CONTROL OF A MODEL TRAIN

In this section fuzzy control and decision making are used for acceleration control of a model train on a circular path. The fuzzy acceleration control system showed a marked improvement over crisp control in smoothness of ride.

Previous software experiences with the applications of fuzzy control to traffic (see Reference 1, Chapter 12) has shown very promising results. In over 625 different simulation runs, it was shown that nearly one-half of the waiting time behind red lights at a single intersection can be saved with fuzzy control over standard techniques such as sensor-based or timed cycle approaches. In a cost (waiting time) comparison between fuzzy control and two standard methods, the fuzzy controller drastically improved on the waiting time, and thereby the cost.

3.5.1 Introduction

This section demonstrates the use of fuzzy logic to control the acceleration of a model train. Velocity is a property which humans do not notice, yet acceleration can cause a passenger discomfort. By selecting a comfortable acceleration and maintaining it as a constant, a smooth increase in velocity and thus a smooth and comfortable ride can be achieved.

The acceleration problem, one of maintaining a constant acceleration and/or deceleration for a moving object (for example, a train), has a fuzzy nature. Fuzzy logic was chosen as a means of control for this problem, the idea being that a constant acceleration needs to be achieved as quickly as possible and then maintained. This is also true of deceleration when bringing the train to a stop. The question was: Can fuzzy logic control do better than a human operator in this area?

The first part of the problem involved the selection of a comfortable acceleration constant, which is obviously different for a model train than a real train. For this work a comfortable acceleration was selected by visual inspection of a water cell being pulled by the model train. The most desirable constant acceleration was indicated by the water level in the water cell maintaining an even surface during increase in the train's velocity. From this visual state the acceleration can be calculated with known time and distance. Once the acceleration was selected, the more interesting problem of control was explored. The following sections describe the hardware involved in prototyping the problem and solution.

3.5.2 Model Train Layout

The model train travels on a circular track. A circle is used because any angular momentum experienced by the train becomes uniform in a circle. An oval shape would contribute angular momentum only at the curved ends of the oval. The circle is three feet in diameter from center of track to center of track. While circling, the train pushes an arm which turns an optical encoder at the center of the track circle. The optical encoder generates a number of pulses per revolution. These pulses are fed to a computer which calculates the velocity and acceleration of the train. The arm and optical encoder eliminate the need for attaching wires directly to the moving locomotive; thus, there is no need for an electrical umbilical cord which could become tangled during running of the train. Figure 3.26 shows the layout of the train track, train, arm, and optical encoder. The circle represents the train track; the dot at the center is the optical encoder; the gray rectangle is the train locomotive; and the line between the locomotive and optical encoder is the arm. The model train locomotive pulls a gondola car which has a water cell mounted inside (Figure 3.27). The gondola car has a lead weight mounted to the bottom of it between the wheel bases. This weight helps balance the weight of the water and water cell.

The computer takes the difference of the selected and current accelerations and uses it as an input to the fuzzy controller. The output of the fuzzy controller is an amount of change in acceleration, which might be an increase, a decrease, or nothing. This actually results in the computer increasing, decreasing, or not changing the power to the train. Deceleration follows the same scenario, except that the selected value is negative.

3.5.3 Computer Input

The movement of the train pushing the arm turns the optical encoder. The rotation of the optical encoder generates digital pulses at the rate of 128 pulses per revolution, or one pulse every 2.8 degrees. The pulses are accumulated in a 6-bit counter which eliminates the need for the computer to be constantly reading pulses. Also, since the pulses are digital, there is no need for an analog-to-digital conversion. The 6-bit counter is constructed of two cascading 4-bit counters. Through experimentation, it was determined that one 4-bit counter was insufficient for counting pulses when the train was moving at full velocity. The train's top velocity generates more than 16 pulses per read (strobe of the game and serial ports) by the computer. The 6-bit counter allows for a maximum of 64 pulses to be accumulated.

The 6-bit counter is connected to the computer through the game port (four lines used) and the serial port (two lines used). This provides six lines for reading the 6-bit value from the counter. Use of the game and serial ports simplifies the interface between the computer and the input from the train. No communications protocol is necessary.

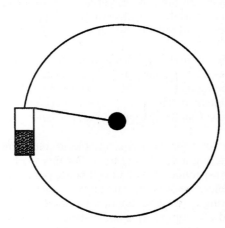

Figure 3.26. A layout of the model train on a circular path.

Hardware Applications of Fuzzy Logic Control

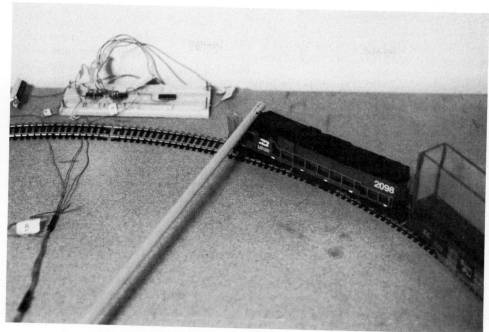

Figure 3.27. Experimental setup of the model train acceleration.

3.5.4 Fuzzy Controller

The fuzzy controller consists of one input and one output. The input to the controller is the difference in the selected acceleration and the train's current acceleration. The membership functions for diff (the input) are Very Negative, Negative, Zero, Positive, and Very Positive (VN, N, Z, P, and VP, respectively). The output from the controller is an amount of change in power. The membership functions for this change (the output) are Large Decrease, Small Decrease, Zero, Small Increase, and Large Increase (LD, SD, Z, SI, and LI, respectively), were chosen in standard fashion as other applications in this chapter. The fuzzy rule-base mapping the input to the output is shown in Figure 3.28. The number of rules is determined by the number of membership functions in the input, which is a total of five. These membership functions and rules will work for deceleration as well because the selected deceleration value will be the negative of the selected acceleration value.

```
IF diff is Z THEN change is Z
IF diff is P THEN change is SI
IF diff is VP THEN change is LI
IF diff is N THEN change is SD
IF diff is VN THEN change is LD
```

Figure 3.28. Fuzzy rules for the train acceleration control system.

3.5.5 Computer Output

The output from the fuzzy controller (change in power) is added to the current power being output to the train. This range of power (in volts) is represented by the computer as a numerical range between zero and 255. The computer outputs a value in this range via the parallel port to a digital-to-analog (D/A) chip. Use of the parallel port for output from the computer again simplifies the interface and avoids the need for communications protocols. The D/A chip converts the eight input lines to one analog output line. The output from the chip is a voltage in the range of zero to 5 volts and is connected to a power regulator circuit. A 15-volt transformer serves as the power supply for the train. This supply is varied by the power regulator circuit as directed by the D/A chip output. The power regulator circuit has an output (to the train track) in the range of 4 to 10 volts, along with an increased current to drive the train. Experimentation showed that 4 volts is just below the amount needed to keep the train moving.

3.5.6 Software Interface and Results

The software interface consists of three parts: the input module, the fuzzy controller module, and the overall driver program. The input module and driver program are written in C compiled with the Turbo C compiler. The fuzzy controller module is written in Togai InfraLogic fuzzy source code and compiled into C with the Togai Fuzzy-C Compiler.[3] All these programs were executed on a personal computer. The input module read the data lines from the counter on the game and serial ports and built the 6-bit value representing the number of pulses. A driver program was written in C to go with the Togai Fuzzy-C language code. The driver program initialized everything, computed the acceleration, called the fuzzy controller, and wrote the output to the parallel port.

The results are heavily based upon visual inspection of the water cell and observation of the train itself. The hope was that the surface of the water in the water cell would angle "uphill" as the train increased acceleration. When a constant acceleration is achieved the water should level itself, and when deceleration begins the water level should angle "downhill." If this state actually occurred, however, the angle was so small it could not be observed. Perhaps the maximum velocity was not large enough. Because of this it was also difficult to select an acceleration for the model train which represented a "comfortable" (for humans) acceleration in real trains. The time it takes the model train to accelerate to maximum is so short that there is not much to observe.

Two methods of acceleration increase were observed, fuzzy controlled and crisp proportional control. The crisp control method is analogous to setting the "throttle" wide open: the train was observed making a "jackrabbit" start and the water in the water cell showed erratic movement. This is analogous to driving a car while holding a cup of coffee: the coffee sloshes around or even spills when the car is abruptly stopped or started. When acceleration was controlled with the fuzzy controller, the train's behavior was much less abrupt. Some movement in the water cell was observed, but the water surface was smoother than that observed with crisp control.

3.5.7 Problems and Limitations

The main problem with this experiment is one of granularity. The computer measures its own internal time in clicks, with 18.2 clicks per second. This calculates to be approximately 0.05 seconds per click. In 0.05 seconds at full speed the train travels approximately two pulses. More accurate accelerations could be calculated with more accurate timing of the pulses, which would require a computer whose internal clock generates more clicks per second.

Although the lack of communications protocol simplifies the design of the system, it also causes a potential problem. This problem occurs with reading the values of the counter through the game and serial ports. It is possible for the counter to receive a pulse during the time the computer is reading from those ports. This could cause an erroneous value to be input to the computer. Hence, a check was added to identify and discard any calculated velocities which appeared out of range (faster than the observed maximum velocity of the train).

Another limitation involved the power regulator circuit. Although the power supply offered a maximum of 15 volts, the power regulator circuit was able to vary a range of only 4 to 10 volts. An improved power regulator circuit could obtain smoother control of the train and allow the train to be accelerated over a longer period of time, allowing the fuzzy control to have a greater effect.

3.6 FUZZY CONTROL OF A THERMOELECTRIC DEVICE-BASED REFRIGERATION SYSTEM

3.6.1 Introduction

This section deals with two important issues in manufacturing: (1) Environmentally Conscious Manufacturing (ECM); and (2) fuzzy logic as an alternative method of controlling manufacturing processes, thereby creating a new type of so-called *smart manufacturing products*. The specific application presented here is a refrigeration system which is not based on Chlorofluorocarbon (CFC) or any other chemical substance, but rather on electronic refrigeration using thermoelectric devices (TEDs).[13, 14]

Moreover, this section describes the potential of fuzzy logic for system control on the lowest level. Since the chamber temperature is a first-order linear system, its control is easily achieved via a PI linear controller. However, knowing that a TED is a dynamic device, and lacking the exact model of the system, an FLC (fuzzy logic controller) is proposed as a possible solution for temperature control.[15, 16]

3.6.2 Thermoelectric Devices

In simple terms, a thermoelectric device (or TED) is a heat pump that transports heat from one location to another. This phenomenon is traceable to the early work of Seebeck and Peltier as reported in,[14] whose work led to the contemporary thermocouple and thermoelectric devices. Peltier's original junction would generate only small temperature changes, while contemporary devices can achieve drops as large as 150°C below ambient temperature. The reason for this performance improvement in the contemporary thermoelectric device lies in the high-technology materials used in its fabrication. The selection of the material used in such a cell is based mainly on operational requirements, and can range from antimony, lead, or bismuth all the way to telluride. Figure 3.29 illustrates the construction of a simple single-junction TED. Each TED contains anywhere from three to 127 such junctions, electrically connected in series, but thermally in parallel. When the electric current is reversed, so is the cool/heat function of the device.

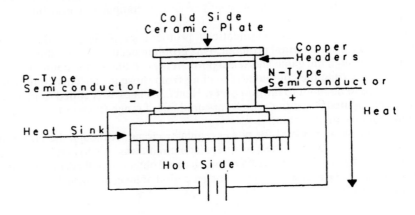

Figure 3.29. Simple single-junction TED.

Thermoelectric cooling can be accomplished only if the generated heat is rapidly transported from the junction; otherwise, a transmigration of thermal energy back to the cold side occurs, reversing the previous cooling effect. There are additional drawbacks to using TEDs. Most vital among these is low-energy efficiency—when electrical input energy is substantially higher than thermal output derived. This problem is of significant concern when applying thermoelectric modules to cooling applications. In fact, many companies stopped using TEDs in this application years ago because of their low overall efficiency. Part of the problem in using TEDs is that during normal operation, heat builds up between the hot and cold plates, affecting the module's performance and efficiency. Figures 3.30-3.32 show the current response and inter-ceramic response of such a device to the step input of 12 volts, as well as the device's resistance.

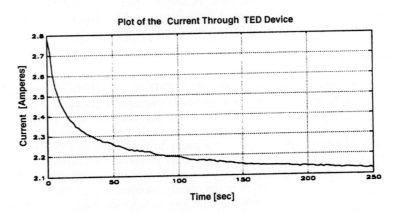

Figure 3.30. Current through a TED.

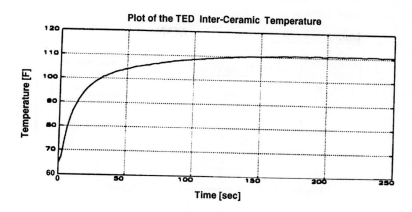

Figure 3.31 Inter-ceramic temperature of a TED.

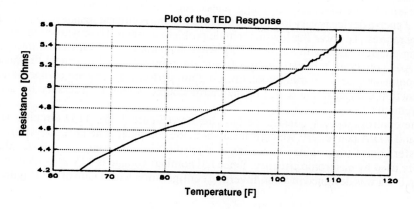

Figure 3.32 Dynamic resistance of a TED.

It can be seen that a TED is a dynamic device whose resistance depends on the inter-ceramic temperature. Consequently, lowering inter-ceramic temperature will cause a decrease in the TED's resistance and, therefore, an increase in TED current for the same step-input voltage. This increase in current means that TED efficiency could improve drastically, making the TED a valid alternative energy source for refrigeration.

3.6.3 Real-Time System Set-Up

In order to perform a real-time experiment, a TED was set up as shown in Figure 3.33 with control circuitry providing the voltage to the TED, while the measurement circuitry provided a means of measuring the temperature of the water in the reservoir.

leaving the computer, is actually a software controller based on classical control theory or fuzzy logic theory.

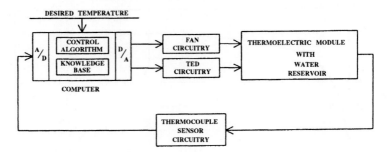

Figure 3.33. Block diagram of system set-up.

The measurement circuitry consists of thermocouple sensor circuitry (Figure 3.34) and an A/D converter[17] which is installed in the computer. Measurement circuitry uses a Type K thermocouple as a transducer transferring temperature into the voltage. A thermocouple was chosen because of its low cost, raggedness, and ease of implementation; however, the output of such a device is very low in magnitude and has a nonlinear response over the range of temperatures. To solve the low-magnitude problem, op-amp U1-c was set up in a non-inverting configuration with a gain of 300. After this voltage was transferred into a count via an A/D converter, a linearization software program was used to cancel the measurement-nonlinearity error. Furthermore, this program performs a software compensation for cold-junction compensation (CJC) of the thermocouple, which is necessary for accurate measurements.

The control circuitry consists of fan circuitry (Figure 3.35), TED circuitry (Figure 3.36), and a D/A converter[17] which is installed in the computer. The D/A converter is used to transfer counts (determined by a control algorithm) into the range of voltages that can be easily used by electronic circuitry. Since fan nominal voltage is 12V at 0.5A of current, op-amp U1-a was designed in a non-inverting configuration with a gain of 1.2 (Figure 3.36)

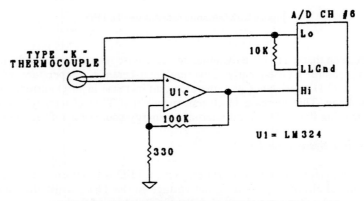

Figure 3.34. Thermocouple measurement circuitry.

such that the input range of 0 to 10V is now transferred to a range of 0 to 12V. Op-amp U2-a is configured as a voltage follower, but is used as a motor driver because of its high current capabilities. Capacitor C_1 is used to filter out the noise produced by a fan (DC motor M_1). The purpose of the fan circuit is rapid heat removal from the heat sink, which is thermally connected to the hot plate of the TED. This action is necessary to improve TED efficiency.

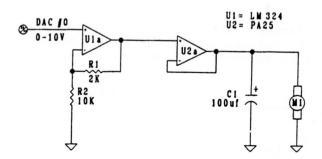

Figure 3.35 Fan circuitry.

The circuit in Figure 3.36 is a linear amplifier for a TED. Op-amp U1-b is configured as a non-inverting amplifier with a gain of 2.7. This transfers the D/A output range of 0 to 5V into a range of 0 to 13.5V. However, Darlington pair Q_1 lowers this voltage by 1.5 volts (each V_{BE} is 0.75 volts), making the TED module range 0 to 12V. Therefore, 5V at the output of DAC #1 corresponds to 12V and thus provides the maximum cooling effect on the TED.

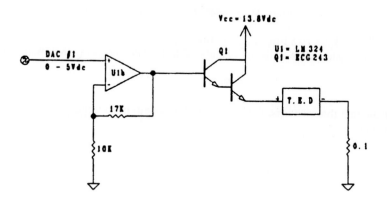

Figure 3.36 TED circuitry.

3.6.4 Proportional-Integral Controller Design

In Close and Frederick,[16] a dynamic model of a thermal system was developed with transfer function

$$G_p(s) = \frac{\frac{1}{C}}{s + \frac{1}{RC}} \tag{3.5}$$

where R is thermal resistivity (conduction resistance between surroundings), and C is thermal capacitance (stored heat). Equation (3.5) is supported by Figure 3.37, which shows the system's open-loop response to a step input (classical first-order response).

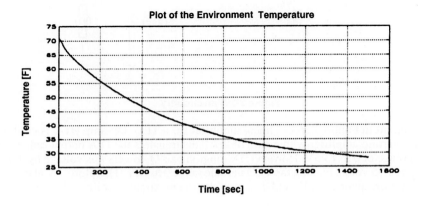

Figure 3.37. Open-loop response to the step input.

From Figure 3.38, rise time–i.e., the time the system takes to change 10% to 90% of initial value to steady-state value–is calculated to be 1000 seconds. Knowing that

$$t = RC \tag{3.6}$$

and using the fact that rise time for the first order system is governed by T_R @ 2, equation (3.5) becomes

$$\cdot G_p(s) = \frac{\frac{1}{C}}{s + 0.002} \tag{3.7}$$

Let $K=1/C$, and apply steady-state theorem to (3.3) with $R(s) = 12/s$ and $ess=28$. Equation (3.7) becomes

$$\cdot G_p(s) = \frac{0.00467}{s + 0.002} \tag{3.8}$$

However, using equation (3.8), the PI controller is governed by the following equation:

$$G_c(s) = \frac{K_p\left(s + \frac{K_I}{K_p}\right)}{s} \qquad (3.9)$$

which implies that the controller will add to the system *Pole* at zero and *Zero* at the value of K_I/K_p. Therefore, the constraints in designating the values for K_I and K_p are:

1) $0 > K_I/K_p > 0.02$
2) $K_p = 1.9034$

The implementation algorithm for a PI controller was derived using Figure 3.40 as follows:

$u(k) = K_p e(k) + K_I \{T_s e(k) + 0.5 T_s [e(k-1) - e(k)]\}$

or

$u(k) = K_p e(k) + 0.5 K_I T_s [e(k-1) + e(k)]$.   (3.10)

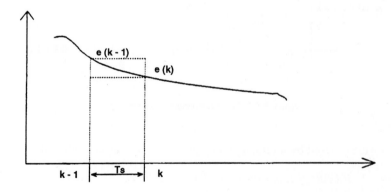

Figure 3.38. Error representation.

3.6.5 Fuzzy Logic Controller Design

Using the concepts of fuzzy membership functions and "expert" knowledge of TED behavior, it was decided to govern the FLC by magnitude and direction of error. Values of these variables determine the sign and the magnitude of the voltage change on the TED. Membership functions are shown in Figures 3.40 and 3.41.

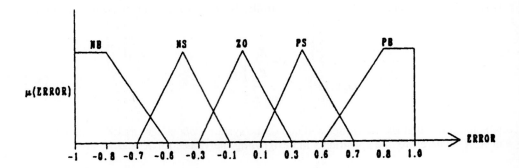

Figure 3.39. Error membership function.

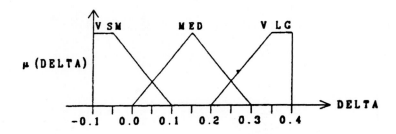

Figure 3.40. Delta membership function.

Fuzzy rules were devised by intuition and by previous experience with TEDs, as follows:

    Rule 1: *If ERROR is PB then DELTA is VLARGE.*
    Rule 2: *If ERROR is PS then DELTA is MED.*
    Rule 3: *If ERROR is ZO then DELTA is VSMALL.*
    Rule 4: *If ERROR is NS then DELTA is MED.*
    Rule 5: *If ERROR is NB then DELTA is VLARGE.*

Using a Fuzzy-C Compiler,[3] membership functions and rules were then compiled into a subroutine called up by the main control algorithm. More detail inforamtion on this application can be obtained in refefence.[19]

3.6.6 Experiment and Results

Data for Figures 3.41 and 3.42 were achieved in real-time, and were graphed using Matlab[18] software. Error calculations were determined as a measured temperature versus

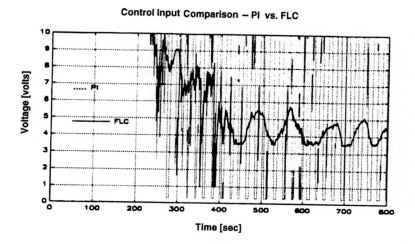

Figure 3.41 Control input - PI vs. FLC.

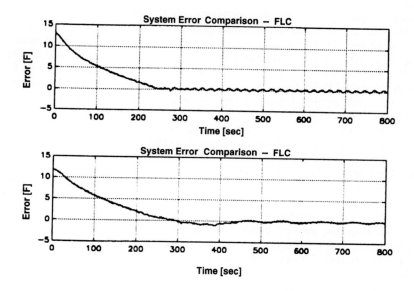

Figure 3.42. System error - PI vs. FLC.

desired temperature entered by the user.

Figure 3.43 shows that both the PI and the FLC show similar results for system error. Even though the FLC has a bigger undershoot, its steady-state error is smaller than that of the PI controller, which makes these compensators comparable in error performance. For a PI controller to achieve such a performance, its control input (shown in Figure 3.41) was fluctuating drastically from 0 to 10V, with an average of 5V. The FLC, on the other hand, had a much smoother control curve, with an average of about 4 volts. Since criteria for the TED controller design require an energy-efficient system, the FLC's energy savings of 20 percent make it an obvious choice.

In this experiment, performance of a fuzzy logic controller was compared to the performance of a conventional PI controller. A mathematical model of the system, though not available, was achieved through system identification techniques. This mathematical model is a necessary ingredient in the design of a conventional controller. On the other hand, when implemented on the lowest level, the FLC does not need a mathematical model but does need the availability of a human expert who can specify the rules underlying the system's behavior.

3.7 CONCLUSION

This chapter presents a few hardware applications of fuzzy logic and intelligent control at the University of New Mexico's CAD Laboratory for Intelligent and Robotic Systems. The following are some individual conclusions on the projects described in this chapter.

By implementing fuzzy control in the laser beam tracking system, overshoot was significantly reduced and limit cycling was virtually eliminated. However, justification of the added complexity of the micro-controller cannot be accomplished until a comparison is made with conventional proportional-plus-integral control. These results simply show the ease of applying fuzzy control to a problem without using a mathematical model of the system. Future study should be considered in modeling the system and implementing a PI controller to further establish or dispute the viability of fuzzy control for this particular problem. Additional possibilities include reducing the amount of hardware. The error electronics and nonlinear motor driver circuit can be removed and replaced by the fuzzy controller.

The power generation systems case study showed that fuzzy logic can be used in power systems control. The main result, however, is that this method provides a simple means of designing a controller for a nonlinear system without complicated calculations and expensive hardware. In some cases, the fuzzy controller was shown to provide marked improvements over non-fuzzy controllers in terms of overshoot suppression and speed of response.

In the robot control applications, the fuzzy controller was able to suppress the robot trajectory overshoot and perform better than traditional PD control under all conditions tested. Furthermore, one set of fuzzy membership functions was sufficient to accommodate the variations that occurred in operating and controlling the different links of the robot. This implies that fuzzy control is robust and able to adapt to many unforeseen elements inevitable in any practical implementation.

Hardware implementation of the fuzzy control of the model train does offer an improved method of controlling acceleration. The desired amount of improvement was not achieved, but improvement over an "open throttle" was observed. It should be noted that in this hardware experiment, no feedback was used from the surface of the liquid on the trailing wagon behind the locomotive unit, and yet a smoother movement was demonstrated from rest to maximum velocity and back to rest. The small scale of this experiment affected what could be observed, and the problems discussed in the previous section dampened the potential results. Further refinements of this implementation of the fuzzy acceleration control for a model train are underway.

Fuzzy control of the thermoelectric refrigeration system showed that a marked improvement can be obtained in the settling time of the responses. The temperature of the chamber varied in an oscillatory mode using crisp control, while no significant sustained oscillation could be observed in the fuzzy logic case.

REFERENCES

1. Jamshidi, M., N. Vadiee and T. Ross (eds.) "Fuzzy Logic and Control: Hardware and Software Applications." *PH Series on Environmental and Intelligent Manufacturing Systems* (M. Jamshidi, Series Editor), Vol. 2. Prentice Hall Publishing Company, Englewood Cliffs, NJ (1993).

2. Zadeh, L. A. *Fuzzy Sets, Information and Control.* Vol. 8 (1965), pp. 335-353.

3. Togai InfraLogic, Inc. *Fuzzy-C-Development Systems User's Guide.* Irvine, CA (1990).

4. Drier, M. *FULDEK - FUzzy Logic DEvelopment Kit User's Guide.* Bell Helicopter Textron, Inc., Fort Worth, TX (1992).

5. NeuraLogix, Inc. *Fuzzy Microcontroller NLX-230 User's Guide.* Sanford, FL (1991).

6. Hiyama, T., and C. M. Lim. "Application of Fuzzy Logic Control Scheme for Stability Enhancement of a Power System." *IFAC Symposium on Power Systems and Power Plant Control* (Aug. 1989), Singapore.

7. Hsu, Y.Y. and C.H. Cheng. "Design of Fuzzy Power System Stabilisers for Mulitmachine Power Systems." IEE Proceedings, May 1990, Vol. 137, pp. 233-238.

8. Hassan, M.A.M., O.P. Malik, and G.S. Hope. "A Fuzzy Logic Based Stabilizer for a Synchronous Machine." *IEEE Trans. on Energy Conversion* (Sept. 1991), pp. 407-413.

9. David, A. K., and Rongda, Z., "An Expert System with Fuzzy Sets for Optimal Planning." *IEEE Trans. on Power Systems* (Feb. 1991), pp. 59-65.

10. Tomsovic, K. "A Fuzzy Linear Programming Approach to the Reactive Power/Voltage Control Problem." *IEEE Trans. on Power Systems* (Feb. 1992), pp. 287-293.

11. Craig, John J. *Introduction to Robotics.* Addison-Wesley, Reading, Mass. (1989).

12. Larsen, D. G., P. R. Rony, J. A. Titus, and C. A. Titus. *Interfacing and Scientific Data Communications Experiments.* Howard W. Sams & Co., Inc., Indianapolis, Indiana (1979).

13. ITI FerroTec. *Thermoelectric Product Catalog and Technical Reference Manual.* Chelmsford, MA. 1992.

14. O'Geary, D. and T. O'Geary. "Development of a Reliable, Efficient Non-CFC Based A/C System." Proc. Environmental Conscious Manufacturing, Santa Fe, NM (1991), pp. 183-191.

15. Rogers, S. "Infinite Band Controller with a Fuzzy Tuner." ISA Trans., Vol. 31, No. 4 (1992), pp. 19-24.

16. Close, C. M. and D. K. Frederick. *Modeling and Analysis of Dynamic Systems.* Houghton Mifflin Company, New York (1978).

17. Phillips, C. L. and R. D. Harbor. *Feedback Control Systems.* Prentice-Hall, Englewood Cliffs, NJ (1988).

18. Mathworks Inc. *Matlab for Windows User's Guide.* Natick, MA., December 1991.

19. D. Barak. "Real-Time Fuzzy Logic-Based Control of Industrial Systems." MS Thesis, CAD Laboratory for Intelligent and Robotic Systems, Dept. EECE, University of New Mexico, Albuquerque, NM (May, 1993).

APPENDIX 3-A
Togai InfraLogic Fuzzy-C .TIL
program for the robot control problem

```
/* Two-link Robot Control using Fuzzy Logic.
   Membership functions.      */

PROJECT robot_control

VAR e_theta1
    TYPE float
    MIN -3.15
    MAX 3.15

  MEMBER ZE
        POINTS -0.5 0 0 1 0.5 0
    END
    MEMBER PS
        POINTS 0 0 0.5 1 1 0
    END
    MEMBER NS
        POINTS 0.0 0 -0.5 1 -1 0
    END
    MEMBER PM
        POINTS 0.5 0 1 1 3.15 1
    END
    MEMBER NM
        POINTS -0.5 0 -1 1 -3.15 1
    END
END
VAR e_theta2
    TYPE float
    MIN -3.15
    MAX 3.15

  MEMBER ZE
        POINTS -0.5 0 0 1 0.5 0
    END
    MEMBER PS
        POINTS 0 0 0.5 1 1 0
    END
    MEMBER NS
        POINTS 0.0 0 -0.5 1 -1 0
    END
    MEMBER PM
        POINTS 0.5 0 1 1 3.15 1
```

```
            END
            MEMBER NM
                    POINTS -0.5 0 -1 1 -3.15 1
            END
    END
    VAR dtheta1
        TYPE float
        MIN -5.0
        MAX 5.0
        MEMBER ZE
                POINTS -1 0 0 1 1 0
        END
        MEMBER PS
                POINTS 0 0 1 1 5 1
        END
        MEMBER NS
                POINTS -5 1 -1 1 0 0
        END
    END
    VAR dtheta2
        TYPE float
        MIN -5.0
        MAX 5.0
        MEMBER ZE
                POINTS -1 0 0 1 1 0
        END
        MEMBER PS
                POINTS 0 0 1 1 5 1
        END
        MEMBER NS
                POINTS -5 1 -1 1 0 0
        END
    END
    VAR torque1
        TYPE float
        MIN -2000
        MAX 2000
        MEMBER ZE
                POINTS -200 0 0 1 200 0
        END
        MEMBER PS
                POINTS 0 0 200 1 400 0
        END
        MEMBER NS
                POINTS -400 0 -200 1 0 0
        END
        MEMBER PM
                POINTS 200 0 400 1 600 0
```

```
            END
        MEMBER NM
                POINTS -600 0 -400 1 -200 0
        END
        MEMBER PL
                POINTS 400 0 600 1 2000 1
        END
        MEMBER NL
                POINTS -2000 1 -600 1 -400 0
        END
    END
    VAR torque2
        TYPE float
        MIN -2000
        MAX 2000
        MEMBER ZE
                POINTS -200 0 0 1 200 0
        END
        MEMBER PS
                POINTS 0 0 200 1 400 0
        END
        MEMBER NS
                POINTS -400 0 -200 1 0 0
        END
        MEMBER PM
                POINTS 200 0 400 1 600 0
        END
        MEMBER NM
                POINTS -600 0 -400 1 -200 0
        END
        MEMBER PL
                POINTS 400 0 600 1 2000 1
        END
        MEMBER NL
                POINTS -2000 1 -600 1 -400 0
        END
    END
    FUZZY control_rules
        RULE rule1 IF e_theta1 IS NM AND dtheta1 IS PS THEN
                torque1 IS NL END
        RULE rule2
                IF e_theta1 IS NM AND dtheta1 IS ZE THEN
                                                             torque1 IS NM
        END
        RULE rule3
                IF e_theta1 IS NM AND dtheta1 IS NS THEN
                                                             torque1 IS NS
        END
```

RULE rule4
    IF e_theta1 IS NS AND dtheta1 IS PS THEN
    torque1 IS NM
END
RULE rule5
    IF e_theta1 IS NS AND dtheta1 IS ZE THEN
    torque1 IS NS
END
RULE rule6
    IF e_theta1 IS NS AND dtheta1 IS NS THEN
    torque1 IS ZE
END
RULE rule7
    IF e_theta1 IS ZE AND dtheta1 IS PS THEN
    torque1 IS NS
END
RULE rule8
    IF e_theta1 IS ZE AND dtheta1 IS ZE THEN
    torque1 IS ZE
END
RULE rule9
    IF e_theta1 IS ZE AND dtheta1 IS NS THEN
    torque1 IS PS
END
RULE rule10
    IF e_theta1 IS PS AND dtheta1 IS PS THEN
    torque1 IS ZE
END
RULE rule11
    IF e_theta1 IS PS AND dtheta1 IS ZE THEN
    torque1 IS PS
END
RULE rule12
    IF e_theta1 IS PS AND dtheta1 IS NS THEN
    torque1 IS PM
END
RULE rule13
    IF e_theta1 IS PM AND dtheta1 IS PS THEN
    torque1 IS PS
END
RULE rule14
    IF e_theta1 IS PM AND dtheta1 IS ZE THEN
    torque1 IS PM
END
RULE rule15
    IF e_theta1 IS PM AND dtheta1 IS NS THEN
    torque1 IS PL
END

RULE rule16
    IF e_theta2 IS NM AND dtheta2 IS PS THEN
    torque2 IS NL
END
RULE rule17
    IF e_theta2 IS NM AND dtheta2 IS ZE THEN
    torque2 IS NM
END
RULE rule18
    IF e_theta2 IS NM AND dtheta2 IS NS THEN
    torque2 IS NS
END
RULE rule19
    IF e_theta2 IS NS AND dtheta2 IS PS THEN
    torque2 IS NM
END
RULE rule20
    IF e_theta2 IS NS AND dtheta2 IS ZE THEN
    torque2 IS NS
END

RULE rule21
    IF e_theta2 IS NS AND dtheta2 IS NS THEN
    torque2 IS ZE
END

RULE rule22
    IF e_theta2 IS ZE AND dtheta2 IS PS THEN
    torque2 IS NS
END
RULE rule23
    IF e_theta2 IS ZE AND dtheta2 IS ZE THEN
    torque2 IS ZE
END
RULE rule24
    IF e_theta2 IS ZE AND dtheta2 IS NS THEN
    torque2 IS PS
END
RULE rule25
    IF e_theta2 IS PS AND dtheta2 IS PS THEN
    torque2 IS ZE
END
RULE rule26
    IF e_theta2 IS PS AND dtheta2 IS ZE THEN
    torque2 IS PS
END
RULE rule27

                IF e_theta2 IS PS AND dtheta2 IS NS THEN
                                                                torque2 IS PM
       END
       RULE rule28
                IF e_theta2 IS PM AND dtheta2 IS PS THEN
                                                                torque2 IS PS
       END
       RULE rule29
                IF e_theta2 IS PM AND dtheta2 IS ZE THEN
                                                                torque2 IS PM
       END
       RULE rule30
                IF e_theta2 IS PM AND dtheta2 IS NS THEN
                                                                torque2 IS PL
END
END
/* The following CONNECT Objects specify that e_theta1, e_theta2, dtheta1 and dtheta2 are the inputs to the Control_rules knowledge base while torque1 and torque2 are the outputs from the Control_rules. */

CONNECT
    FROM e_theta1
    TO control_rules
END
CONNECT
    FROM e_theta2
    TO control_rules
END
CONNECT
    FROM dtheta1
    TO control_rules
END
CONNECT
    FROM dtheta2
    TO control_rules
END
CONNECT
    FROM control_rules
    TO torque1
END
CONNECT
    FROM control_rules
    TO torque2
END
END

APPENDIX 3-B
THE FUZZY RULE BASE FOR THE CONTROL OF THE TED SYSTEM

FUZZY RULES

Rule 1: IF DF is Z AND DVAC is Z
THEN DIshunt is Z AND DVf is Z

Rule 2: IF DF is Z AND DVAC is PS
THEN DIshunt is Z AND DVf is PS

Rule 3: IF DF is Z AND DVAC is PM
THEN DIshunt is Z AND DVf is PM

Rule 4: IF DF is Z AND DVAC is PB
THEN DIshunt is PS AND DVf is PB

Rule 5: IF DF is Z AND DVAC is NS
THEN DIshunt is Z AND DVf is NS

Rule 6: IF DF is Z AND DVAC is NM
THEN DIshunt is Z AND DVf is NM

Rule 7: IF DF is Z AND DVAC is NB
THEN DIshunt is NS AND DVf is NB

Rule 8: IF DF is NS AND DVAC is Z
THEN DIshunt is NS AND DVf is Z

Rulc 9: IF DF i8 NS AND DVAC is PS
THEN DIshunt is NS AND DVf is PS

Rule 10: IF DF is NS AND DVAC is PM
THEN DIshunt is NS AND DVf is PM

Rule 11: IF DF is NS AND DVAC i8 PB
THEN DIshunt is Z AND DVf is PB

Rule 12: IF DF is NS AND DVAC is NS
THEN DIshunt i8 NS AND DVf is NS

Rule 13: IF DF is NS AND DVAC is NM
THEN DIshunt is NS AND DVf is NM

Rule 14: IF DF is NS AND DVAC is NB
THEN DIshunt is NS AND DVf is NB

Rule 15: IF DF is PS AND DVAC is Z
THEN DIshunt is PS AND DVf is Z

Rule 16: IF DF is PS AND DVAC is ISPS
THEN DIshunt is PS AND DVf is PS

Rule 17: IF DF is PS AND DVAC is PM
THEN DIshunt is PS AND DVf is PM

Rule 18: IF DF is PS AND DVAC is PB
THEN DIshunt is PS AND DVf is PB

Rule 19: IF DF is PS AND DVAC is NS
THEN DIshunt is PS AND DVf is NS

Rule 20: IF DF is PS AND DVAC is NM
THEN DIshunt is PS AND DVf is NM

Rule 21: IF DF is PS AND DVAC is NB
THEN DIshunt is Z AND DVf is NB

Rule 22: IF DF is NM AND DVAC is Z
THEN DIshunt is NM AND DVf is PS

Rule 23: IF DF is NM AND DVAC is PS
THEN DIshunt is NM AND DVf is PM

Rule 25: IF DF isNM AND DVAC is PB
THEN DIshunt is NM AND DVf is PB

Rule 26: IF DF is NM AND DVAC is NS
TNEN DIshunt is NM AND DVf is Z

Ruk 27: IF DP is NM AND DVAC is NM
THEN DIshunt is NM AND DVf is NS

Rule 28: IF DF is NM AND DVAC isNB
THEN DIshunt isNM AND DVf is NB

Rule 29: IF DF is PM AND DVAC is Z
THEN DIshunt is PM AND DVf is NS

Rule 30: IF DF u PM AND DVAC is PS
THEN DIshunt is PM AND DVf is Z

Rule 31: IF DF is PM AND DVAC is PM
THEN DIshunt is PM AND DVf is PS

Rule 32: IF DF is PM AND DVAC isPB
THEN DIshunt is PM AND DVf is PB

Rule 33: IF DF is PM ANI)DVA is NS
THEN DIshunt is PM AND DVf is NM

Rule 34: IF DF is PM AND DVAC is NM
THEN DIshunt is PM AND DVf is NM

Rule 35: IF DF is PM AND DVAC is NB
THEN DIshunt is PM AND DVf is NB

Rule 36: IF DF is NB AND DVAC is Z
THEN DIshunt is NB AND DVf is PS

Rule 37: IF DF is NB AND DVAC is PS
THEN DIshunt is NB AND DVf is PM

Rule 38: IF DF is NB AND DVAC is PM
THEN DIshunt is NB AND DVf is PM

Rule 39: IF DF is NB AND DVAC is PB
THEN DIshunt is NB AND DVf is PB

Rule 40: IF DF is NB AND DVAC is NS
THEN DIshunt is NB AND DVf is Z

Rule 41: IF DF is NB AND DVAC is NM
THEN DIshunt is NB AND DVf is NS

Rule 42: IF DF is NB AND DVAC is NB
THEN DIshunt is NB AND DVf is NB

Rule 43: IF DF is PB AND DVAC is Z
THEN DIshunt is PB AND DVf is NS

Rule 44: IF DF is PB AND DVAC is PS
THEN DIshunt is PB AND DVf is Z

Rule 45: IF DF is PB AND DVAC is PM
THEN DIshunt is PB AND DVf is PS

Rule 46: IF DF is PB AND DVAC is PB
THEN DIshunt is PB AND DVf is PB

Rule 47: IF DF is PB AND DVAC is NS
THEN DIshunt is PB AND DVf is NM

4

FUZZY EXPERT SYSTEMS

R. A. Aliev, Baku University
Baku, AZERBAIJAN

This chapter is devoted to the problem of designing of fuzzy expert systems. The fuzzy expert system ESPLAN, used for resolving the scheduling tasks of the oil refinery plant, will be described as an example. The problems of fuzzy knowledge representation and logical inference will be discussed, as well as some perspective applications of fuzzy expert systems in social management.

4.1 INTRODUCTION

One of the most effective approaches to the control of complex industrial processes is the use of intelligent controllers. We come to this point of view by recognizing that reasonable control of such processes is almost impossible due to the influence of various disturbances. So the use of intelligent control systems gives us a chance to implement the experience and knowledge of a human operator who, in practice, overcomes the above-mentioned technological disturbances. It ought to be mentioned that due to these disturbances complicated technological processes are operating under conditions of uncertainty. And since the nature of this uncertainty is often non-statistical, it may be classified as fuzzy uncertainty, as defined by Zadeh.[2] That is why the main feature of developed control systems is a fuzzy representation of knowledge bases, namely a production rule form of fuzzy knowledge representation.

4.2 THE STRUCTURE OF THE SYSTEM

The common structure of developed intelligent control systems has the next form, shown in Figure 4.1.

This chapter presents some aspects of the shell of expert systems ESPLAN, namely, the problems of fuzzy knowledge representation inthe system's knowledge base and the problem of inference organization. The mathematical description of knowledge in the shell of the expert system ESPLAN is solvedby means of fuzzy interpretation of antecedents and consequents in production rules.

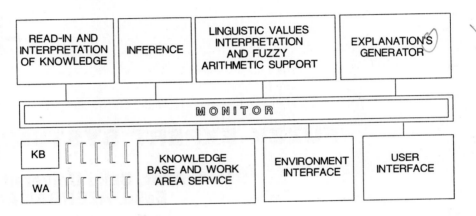

Figure 4.1.

4.3 THE KNOWLEDGE REPRESENTATION

For the knowledge representation in the shell of the expert system ESPLAN was suggested developed form of production rules. The antecedent of the rule contains a conjunction of binary logical connectives as shown in Figure 4.2: <name of object> $\left\{ \begin{array}{c} = \\ \neq \end{array} \right\}$ < linguistic value > named elementary antecedent (for instance, " IF quantity of petroleum = great AND demanded output of residual oil ≠ little AND ..."). The consequent of the rule is a list of imperatives, among which may be some operator-functions (i.e. input and output of objects, values, operations with the segments of a knowledge base, etc.). Each rule may be added by the confidence degree cf ∈ [0,100] and by the author's explanation

Fuzzy Expert Systems

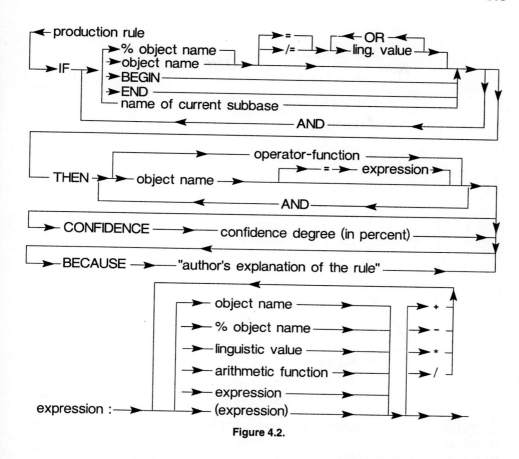

Figure 4.2.

of the rule. For example, let us take a rule from the knowledge base in Figure 4.2:

IF the remainder of local petroleum = great
THEN the load of 17th unit = great
CONFIDENCE 90
BECAUSE "If there is 2 great remainder, then the 17th unit may be completely loaded"

For appropriate interpretation of linguistic values in ESPLAN, the fuzzy sets theory is used. Each linguistic value has a correspondent membership function which is built using the parametric LR-format (Figure 4.3).

Here, each membership function is defined by 4 parameters: α- left deviations, ml- left peak, mr- right peak, β- right deviation, i.e. $\mu_a(u) = (\alpha_a, ml_a, mr_a, \beta_a)$.

So we have the analytic form of the membership function shown in Figure 4.3:

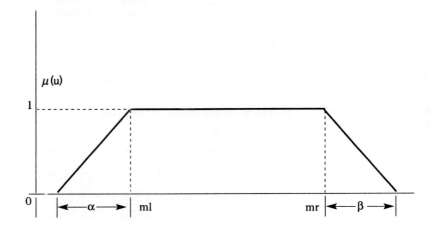

$$\mu(u) = \begin{cases} 1 - \dfrac{ml-u}{\alpha}, & if\ ml-\alpha \le u \le ml \\ 1, & if\ ml \le u \le mr \\ 1 - \dfrac{ml-u}{\alpha}, & if\ mr \le u \le ml+\beta \\ 0, & in\ other\ cases \end{cases}$$

Figure 4.3.

The graphical representation of this membership function is a trapezoid with the tops ml and mr and plumb deviation α and β, respectively. Such representation allows the defining of unimodal membership functions, but the knowledge representation language of ESPLAN also allows the definition of sometimes-polimodal membership functions by using the construction "OR," for instance, "low OR near 50." The result of interpretation is a disjunction (max) of membership functions of composite elements.

The subsystem of fuzzy arithmetic and linguistic values processing provides automatic interpretation of linguistic values like "high," "low," "OK," "near...," "from ... to ..." and so on; i.e. for each linguistic value this subsystem automatically computes α, ml, mr and β using universums of correspondent variable. The user of the system may define new linguistic values, modify built-in values, and explicitly prescribe the membership function in any place where linguistic values are tolerable.

4.4 INFERENCE

The efficiency of an inference engine greatly depends on the knowledge base's internal organisation. That is why ESPLAN realizes a paradigm "network of production rules" similar to a semantic network. Here, the nodes are rules and vertexes are objects.

In brief, the inference mechanism acts as follows. Firstly, some objects take some

Fuzzy Expert Systems

values (initial data). Then, all production rules containing each of these objects in antecedent are chosen from the knowledge base. For these rules the degree of truth is computed (in other words, the system estimates the truth degree of the fact that current values of objects correspond to values fixed in antecedents). If the truth degree exceeds some threshold, then imperatives from consequent are executed. At that time, the same objects as well as a new one take new values and the process continues until the work area contains "active" objects (an "active" object is an untested one).

The assigned value of the object is also added by the number, named confidence degree, which is equal to the truth degree of the rule.

A truth degree of a rule's antecedent is being calculated according to the algorithm :
Let us consider. An antecedent of a rule has the form:

$$\text{IF ... AND } w_i \begin{Bmatrix} = \\ \neq \end{Bmatrix} a_{ji} \text{ AND ... AND } w_k \begin{Bmatrix} = \\ \neq \end{Bmatrix} a_{jk} \text{ AND ...}$$

Confidence degree of the rule $CF \in]0,100]$.

Objects w_i, w_k etc.. have current values of the form (v, cf) in the work area (here v- linguistic value with its membership function, $CF \in [0,100]$ - confidence degree of the value v).

Truth value of k-th elementary antecedent is equal:

$$r_k = Poss(v_k | a_{jk}) * cf_k, \qquad \text{if the sign is "="}$$

and $r_k = (1 - Poss(v_k | a_k)) * cf_k,$ if the sign is "$\neq$"

$$Poss(v | a) = \max_u \min(\mu_v(u), \mu_a(u)) \in [0,1].$$

For the LR-format of membership function:

$$Poss(v|a) = \begin{cases} 1 - \dfrac{ml_v - mr_a}{\alpha_v + \beta_a}, & if\ 0 < ml_v = mr_a < \alpha_v + \beta_a \\ 1, & if\ \max(ml_v, ml_a) \leq \min(mr_v, mr_a) \\ 1 - \dfrac{ml_v - mr_a}{\alpha_v + \beta_a}, & if\ 0 < ml_a = mr_v < \alpha_a + \beta_v \\ 0, & otherwise \end{cases}$$

The truth degree of the rule :

$$R_j = \dfrac{(\min_k r_k) * CF_j}{100}$$

After the inference is over, the user may obtain for each object the list of its values with confidence degrees which are accumulated in the work area. The desirable value of the object may be obtained using one of the developed algorithms:

$$w_i : (v_i^n, cf_i^n), n = \overline{1, S},$$ S - total number of values

Resulting value :

I. Last - v_i^s

II. The value with maximum confidence degree - $v_i^m \big| cf_i^m = \max_n cf_i^n$

III. The value - $\overline{v}_i = \underset{n}{\Lambda} v_i^n * cf_i^n$, or $\overline{v}_i = \underset{n}{V} v_i^n * cf_i^n$

IV. The average value $\overline{v}_i = \dfrac{\sum_n v_i^n * cf_i^n}{\sum_n cf_i^n}$

IF $x_1 = a_1^i$ AND $x_2 = a_2^i$ AND ... THEN $y_1 = b_1^i$ AND $y_2 = b_2^i$ AND ...

IF ... THEN $Y_1 =$ AVRG (y_1) AND $Y_2 =$ AVRG (y_2) AND ...

ESPLAN has the built-in function AVRG, which allows the computation of an average value. This function simplifies the organization of compositional inference with possibility measures. Here, a confidence degree is used as a possibility measure. So the compositional relation is given as a set of production rules like:

IF $x_1 = a_1^j$ AND $x_2 = a_2^j$ AND ... THEN $y_1 = b_1^j$ AND $y_2 = b_2^j$ AND ...,

here j is a number of a rule (similar to the row of the compositional relation matrix). After all these rules are executed (with different truth degree) the next rule (rules) ought to be executed:

IF ... THEN $Y_1 =$ AVRG (y_1) AND $Y_2 =$ AVRG (y_2) AND ....

The approach basing on construction of compositional relations is often used for synthesis of control systems for technological processes. Particularly, it was implemented at the knowledge base for the control of initial oil-refinery unit.

4.5 THE EXPERT SYSTEM FOR CURRENT SCHEDULING OF OIL-REFINERY PRODUCTION

Here we introduce some aspects of system's application for solving the tasks of scheduling oil-refinery production. The process was partitioned on 4 main subprocesses- initial oil-refinery, catalytic cracking, catalytic reforming and petrol compounding.

{Distribution of petroleum among units 15, 16, 17}
....
{Description of objects: shortname, fullname, minimal value, maximal value, unit of measurement}
OB(S_100, "Plan on local petroleum," 10, 30, "thous.tons");
OB(P_106, "Plan on output of residual oil," 1, 3, "thous.tons");
OB(R_lO6, "Real output of residual oil," 1, 3,"thous.tons");
....
OB(IN_42_106,"Input of residual oil at the unit 42," 0.5, 1.5"thous.tons);
....
OB(E_17_106, "Yield efficiency for residual oil at the unit 17")
....
IF
THEN ...
AND P3=(P_106+IN_42_106)/E_17_106
AND P31=TP_17-P3
BECAUSE "Computing necessary load of 17th unit to provide the plan with respect to the residual oil and the load unit 42.
IF P31=Less 0
THEN IN_17_100=TP_17 AND OUT_17_106=IN_17_100*E_17_106
AND R_106=OUT_17 106-IN 42_106
AND DISPLAY ("Power of 17th unit is insufficient for the load
AND DISPLAY ("of the unit 42 and demanded plan on residual oil")
BECAUSE "Necessary load of 17th unit is greater than the top capacity of the unit"

4.6 FUZZY HYPOTHESES GENERATING AND ACCOUNTING SYSTEMS

Using ESPLAN one may construct hypotheses generating and accounting systems. Such system contains the rules:

IF <condition$_j$> THEN X=A$_j$ CONFIDENCE CF$_j$

Here "X=A$_j$" is a hypothesis that the object X takes the value A$_j$ Using; some preliminary information, this system generates elements X=(A$_j$,R$_j$), where R$_j$ is a truth degree of j-th rule. In order to account the hypothesis (i.e. to estimate the truth degree that X takes the value A_o) the recurrent Bayes-Shortliff formula, generalized for the case of fuzzy hypotheses, is used:

$$P_o = O; \quad P_j = P_{j-1} + cf_j * Poss(A_o|A) * \left(1 - \frac{P_j - 1}{100}\right).$$

This formula is realized as a built-in function BS:

IF END THEN P-BS(X,A_o);

These systems were implemented in two problem domains: medical diagnostics and forecasting of social conflicts. In a case of medical diagnostics the structure of knowledge has the form:

IF ... AND $symptom_k$. $\begin{Bmatrix} = \\ \neq \end{Bmatrix}$ $value_{jk}$ AND ...
THEN $diagnosis_l$ = {noS3 CONFIDENCE CF_j; ....

IF END THEN Probability of_$diagnosis_l$ = BS($diagnosis_l$, yes) - BS $diagnosis_l$,no);

There was developed the system for a diagnostics of approximately 40 venous and arterial diseases, concerning its stages.

The distinctive feature of the system is a possibility to describe symptoms using fuzzy values. We have also developed models of forecasting the social conflict and producing current advices. For this purpose approximately 70 factors had been chosen, some of them describe the current situation, others have an influence on the current situation. Basing on these factors, approximately 1500 causal relations were obtained. For example,

"IF illegal activities in Karabah = positive
AND punishment for these activities = negative
THEN illegal activities in Azerbaijan = increase
CONFIDENCE 60;"

Here "positive" and "negative" are linguistic values with membership functions, defined at the range [0,100], with peaks in points 100 and 0, respectively. Another linguistic values like "little", "big", etc., are also used in these models.

The exploitation of the system demonstrated it's sufficiently high adequacy to the real processes. Moreover, one of the most important products of the system is obtained full explanation (i.e. motivation) of the forecasts. This explanation allows to conduct a profound analysis of social processes.

In a case of forecasting of social conflicts the structure of the knowledge base has the following form:[8]

IF ... AND $factor_k$, $\begin{Bmatrix} = \\ \neq \end{Bmatrix}$ $value_{jk}$ AND

...

THEN hypothesis_$factor_i$ = $\begin{Bmatrix} increase \\ decrease \end{Bmatrix}$ CONFIDENCE CF_j ;

IF END THEN forecast_$factor_i$ = BS(hypothesis_$factor_i$,increase)- BS(hypothesis_$factor_i$,decrease);

Another system is theDUET system. This system was developed to provide monitoring

Fuzzy Expert Systems

by any continuous production. However, the use of the system revealed its applicability for diagnostics as well.

The main features of the system are:

- operation in a real-time mode;
- three-level structure of the knowledge base (meta-knowledge, rules, frames);
- the structure of the knowledge representation language contains rules and frames;
- developed knowledge input, and modification system;
- activation of the inference engine on initiative of the expert system.

The next direction our research took was hybrid expert systems. In this way, the NET-EXPERT system was developed.

This system designs local computer networks. It solves the following tasks:

- choice of the hardware for the network user;
- choice of the network topology;
- choice of the transmitting medium;
- choice of the symbol rate;
- choice of the network software.

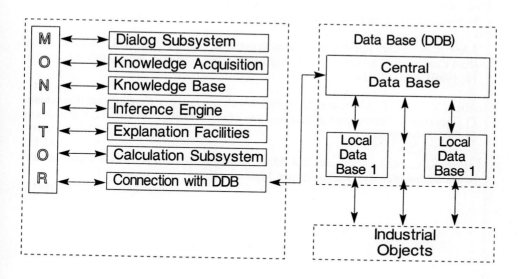

Figure 4.4.

4.7 CONCLUSION

The principles of building fuzzy expert systems described in this chapter, and their implementation in a system such as ESPLAN, makes the creation of a number of applied intelligent systems possible. Their use for various applications, such as manufacturing, social management, etc., shows promise.

REFERENCES

1. Dubois, D., and H. Prade. "Fuzzy Sets and Systems." *Theory and Application.* Academic Press, New York (1980).

2. Dubois, D., and H. Prade. "A typology of fuzzy 'If...then...' rules." James C. Bezdek, ed. *Proc. of 3rd IFSA Congress, "The Coming of Age of Fuzzy Logic."* Seattle, Washington (August 6-11, 1989), pp. 782-785.

3. Shortliffe, E. H. *Computer-Based Medical Consultations: MYCIN.* American Elsevier, New York (1976).

4. Zadeh, L.A. "The concept of a linguistic variable and its application to approximate reasoning." *Part 3, Inform. Sci. 9* (1975), pp. 43-80.

5. Aminzadeh, F., F. S. Wong, and E.H. Ruspini. "A practical view of expert systems for oil exploration: integration of multiple knowledge sources." M. Simann and F. Aminzadeh, eds. *Advances in Geophysical Data Processing, Vol. 3: Artificial Intelligence and Expert Systems in Petroleum Exploration.* Jai Press, Inc., London (1989), pp. 1-17.

6. Aliev, R.A. "Fuzzy intelligent production systems." James C. Bezdek, ed. *Proc. of 3rd IFSA Congress, "The Coming of Age of Fuzzy Logic."* Seattle, Washington (August 6-11, 1989), pp. 311-314.

7. Aliev, R.A., F. T. Aliev, and M. Babaev. *Fuzzy Process Control and Knowledge Engineering in Petrochemical and Robotic Manufacturing.* Verlag TUV Rheinland GmbH, Koln (1991).

8. Aliev, R.A., N.M. Abdikeev, and M.M. Shakhnazarov. *Production Systems with Artificial Intelligence.* Radio i Sviaz, Moscow (1990).

5

PREPROCESSING FUZZY PRODUCTION RULES

M. Schneider and G. Chew, Florida Tech, USA
A. Kandel, University of South Florida, USA
G. Langholz, Tel Aviv University, ISRAEL

Preprocessing is the process of parsing fuzzy production rules within a given knowledge base and establishing the proper relations among the clauses in the fuzzy production rules. In this chapter, we describe the various types of fuzzy production rules and the procedure for establishing relations among those rules. The advantage of this procedure is its simplicity which in turn makes this architecture easily extendible and useful in real time environments.

5.1 INTRODUCTION

Upon completion of the parsing process, an internal structure is created. This internal structure represents the relations among clauses in the knowledge base. The relations represent the similarity between a clause in the conclusion of a rule and a clause in the premise of another rule.

There are several methods to represent the relations among clauses in the knowledge base:

1. Tree structure. In the tree structure representation, a relation between two clauses is represented via two nodes and a link between them.[1] The link represents the similarity measure (or the strength of the similarity) between two clauses.

2. Matrix. In the matrix representation, an entry in the matrix represents the relation (or the similarity measure) between two clauses.[2]

3. Relational list. The relational list (*R-list*) is a list that contains the relations among the rules.[3,4] This list is a 5-tuple list:

$$R\text{-}list = (R_i, C_j, R_k, C_l, M)$$

where R_i and R_k are rule numbers, C_j and C_l are clause numbers and M is the matching factor. The interpretation of each row in the list is: "Clause C_j in the conclusion of rule R_i is similar to clause C_l in the premise of rule R_k with certainty of M".

Regardless of the method of representation, we have to find a method to compute the degree of similarity between two clauses. Also, we have to realize that there are three types of clauses. These are:

1. A regular English clause. For example: **The pressure is high**

2. A mathematical expression. For example: $[X := sqr(Y) + sqrt(Z)/2]$

3. Executable clause. For example: **{Execute P(x,y,z)}**

In the following sections we will show how to parse the various types of clauses and how to find the similarity between any two clauses.

5.2 PROCESSING A REGULAR ENGLISH CLAUSE

The general structure of a clause is "X is Y". A modifier and/or the NOT operator may be inserted to describe "Y". "Y" is associated with two intervals (as will be described later), and they are denoted by $[lb_1, ub_1]$ and $[lb_2, ub_2]$. It is also important to note that "Y" can be one of the following: (1) A word, (2) An interval or (3) a single number. Thus there are four cases to consider:

1. X is Y. In this case, both the NOT and the modifier are off. Thus,

   (a) If "Y" is word then $lb_1 = 1$ and $ub_1 = 1$.

   (b) If "Y" is a single number (N) then $lb_1 = ub_1 = N$.
   (c) If "Y" is an interval [N1,N2] then $lb_1 = N1$ and $ub_1 = N2$.

As can be seen, in all three cases described above, both lb_1 and ub_1 receive a value. The second interval remains undefined.

2. X is MOD Y. Since the modifier modifies "Y", we may apply it to the interval generated in the previous case. Let COMPUTE_RANGE(MOD, NUM,LB,UB) be a procedure that takes as an input the modifier (MOD) and some number (NUM) and generates an interval ([LB,UB]). Assume, for example, that MOD=*more-or-less* and NUM=20. Also assume that *more-or-less* will generate an interval [num-10%,num+10%], thus COMPUTE_RANGE(more-or-less,20,LB,UB) will generate the interval[18,22] such that LB=18 and UB=22. We assign these values to the interval described above. Thus $lb_1 = 18$ and $ub_1 = 22$.

In general, each modifier is associated with some lower bound ($\perp$) and some upper bound ($\top$) values, as we have seen in the example above. The procedure COMPUTE_RANGE searches for the modifier in the modifier list and extracts the values associated with $\perp$ and $\top$. Then it computes the lower bound of the interval ($lb_1 = NUM * \perp$) and the upper bound of the interval ($lb_2 = NUM * \top$). In the above example, if we let $more\_or\_less_\perp = 0.9$ and $more\_or\_less_\top = 1.1$ then according to the example, $lb_1 = more\_or\_less_\perp *NUM = 0.9 * 20 = 18$ and $lb_2 = more\_or\_less_\perp * NUM = 1.1 * 20 = 22$.

The second interval remains undefined.

3. X is NOT Y. In this case we have to negate the interval described earlier. Let α be the smallest possible number in "Y"'s domain and β be the largest number in that domain, then the negation of Y will generate two interval: $[\alpha, lb_1 - \varepsilon]$ and $[ub_1 + \varepsilon, \beta]$ where ε is a small number. So, NOT "Y" generates two distinct intervals where $ub_2 = \beta$, $lb_2 = ub_1 + \varepsilon$, $ub_1 = ub_1 - \varepsilon$ and $lb_1 = \alpha$.

We use ε to guarantee that the intervals generated from negation will not intersect the original interval.

4. X is NOT MOD Y. Here, we first create the proper interval as described in case 1. Then we compute the new interval with the modifier. And finally we apply the NOT operation on the interval created, thus creating two distinct intervals.

We can summarize the above cases in the following tables:

where:

u This indicates that the field is undefined $\top$ is the upper bound of the modifier

Table 5-1. Parsing a clause where Y is a word.

case	Y is a word			
	lb_1	ub_1	lb_2	ub_2
X is Y	1	1	u	u
X is MOD Y	$1 * \bot$	$1 * \top$	u	u
X is NOT Y	0	$1 - \epsilon$	$1 + \epsilon$	2
X is NOT MOD Y	0	$(1 * \bot) - \epsilon$	$(1 * \top) + \epsilon$	2

Table 5-2. Parsing a clause where Y is a number.

case	Y is a number			
	lb_1	ub_1	lb_2	ub_2
X is Y	N	N	u	u
X is MOD Y	$N * \bot$	$N * \top$	u	u
X is NOT Y	α	$N - \epsilon$	$N + \epsilon$	β
X is NOT MOD Y	α	$(N * \bot) - \epsilon$	$(N * \top) + \epsilon$	β

Table 5-3. Parsing a clause where Y is an interval.

case	Y is an interval			
	lb_1	ub_1	lb_2	ub_2
X is Y	N1	N1	u	u
X is MOD Y	$N1 * \bot$	$N2 * \top$	u	u
X is NOT Y	α	$N1 - \epsilon$	$N2 + \epsilon$	β
X is NOT MOD Y	α	$(N1 * \bot) - \epsilon$	$(N2 * \top) + \epsilon$	β

N is a number
N1 is the lower bound of the interval
N2 is the upper bound of the interval
$\bot$ is the lower bound of the modifier

α is the lower bound of Y's domain
β is the upper bound of Y's domain
ϵ is an arbitrary small number

5.3 MATH FUNCTIONS

Expert systems have been designed primarily to capture and process linguistic knowledge in an attempt to model the cognitive processes of the human mind. Fuzzy expert systems, in particular, more closely model these processes by handling uncertain or imprecise linguistic knowledge. Clearly though, real world problems also include those that are more mathematical in nature, and the domain of expert systems should be expanded to include these types of problems. Furthermore, such systems must be able to deal with uncertainty in numeric data. For example, the proper operation of a patient monitoring system depends on processing data gathered from sensors that may be faulty or noisy. Calculations based on this data must maintain as much significance as possible, despite the uncertainty.

Interval arithmetic is an extension of real arithmetic, where the standard operators (+, -, *, /) are applied to intervals of real numbers, rather than to single real numbers. The theory of interval arithmetic was developed as a way to bound the errors inherent in calculations performed on a digital computer.[10] These errors are due to a combination of roundoff, truncation, and limited machine precision.[11] The intervals, which represent the range of values that a variable may take, are carried through all calculations. The result is an interval that exactly bounds the answer.[12]

The relational operators (=, ≠, <, ≤, >, ≥) define logical relations between pairs of numbers or intervals of numbers. This section presents a method for evaluating logical relations between intervals, as an extension to the fuzzy inferencing algorithm.[4-9, 13]

The general structure of a mathematical function is: [EXPRESSION]

There are two types of mathematical expression:[14-15]

1. Logical expression. The logical expression is in the form: $[X <op> Y]$ where X and Y are variables and *op* is some logical relation among the two variables. For example, the relation $[X \leq Y]$ is a legal expression, where X and Y can be single numbers, single intervals, double intervals or any combination of them. The logical expressions can be used only in the premise part of the rules.

2. Assignment statements. The assignment statements are used only in the conclusion part of the rules and have the following form: $[X := \text{Math Expression}]$ For example, the assignment statement $[X := SQR(Y)]$ is a legal assignment statement that computes the square of the variable Y and assigns it to the variable X. Here, Y can be either a single number or an interval.

5.4 EXECUTABLE CLAUSES

The general structure of an executable procedure is:

{*procedure name and parameters*}

When the user wants to link the knowledge base with some executable code, he/she will use the above format. It is a very useful and powerful feature of any expert system. For example:

If Object-1 is not recognized then {RECOGNIZE(object-1,CF)}

In the above example we provide a rule that simply states that if some object is not yet recognized then invoke some procedure "RECOGNIZE" and perform the recognition process on object-1. The certainty of the execution is placed in the variable *CF*.

This implies that the expert system designer must have a set of procedures that are linked to the expert system shell. In order to execute these procedures, a compiler has to be incorporated into the expert system shell. This compiler will parse the request and will link (or call) the proper procedure.

The results from the executable procedure will be placed on the blackboard.

5.5 CREATING THE R-LIST

The Fuzzy Expert Tool (FEST)[3, 4] utilizes the *R-list* to establish links between clauses in the knowledge base. The *R-List* (Relational List) contains the relations among the rules. As was described above, this list is a 5-tuple list:

$$R\text{ - }list = (R_i, C_j, R_k, C_l, M) \tag{5.1}$$

where R_i and R_k are rule numbers, C_j and C_l are clause numbers and M is the matching factor. The interpretation of each row in the list is: "Clause C_j in the conclusion of rule R_i is similar to clause C_l in the premise of rule R_k with certainty of M".

There are several questions regarding the matching process (that is, finding how well two clauses are similar one to another):

1. How to match? We have to develop a technique that will perform the matching process. This technique (or procedure) should reflect our own way of matching two clauses. In other words, the matching procedure should be able to match any two clauses even if the clauses are not identical. The procedure should utilize techniques that will enable us to match clauses which are "almost" identical. The degree of the identity will be measured by some number that will indicate how well two clauses match.

2. What is the certainty of the match? In many cases we do not have a one to one match between two clauses. The result of the matching process should produce a number which will indicate how well two clauses match. This number is called the *matching factor* (*M*). The result of the matching process reflects our confidence in the match. As the result of the matching process increases so is our belief that the two clauses match well, and as the number decreases it should indicate that the two clauses do not match or match poorly. In other words, M=1 indicates a perfect match, and M=0 indicates no match at all. Anything in between indicates the degree in which two clauses match.

Most of the expert systems today do not assign certainty factors dynamically. These expert systems prompt queries about the truth of each clause, and accordingly assign the certainty factors. For example, if the knowledge base contains the rule "If John is between 20 to 30 then John is young", then they may ask:" What is the certainty factor that John is between 20 and 30?" The user may provide a number between 0 and 1 (say, 0.8). If the number is above a certain threshold, then the rule will fire and the conclusion will be "John is young" with the certainty factor of 0.8.

Other expert systems assign the certainty factor to the conclusion but require a complete match between clauses. That means that if the clauses match, then the certainty factor of the conclusion is assigned a priori, and if the clauses do not match, the rule does not fire. For example, we may have the following rule "If John is between 20 to 30 then John is young with CF=0.9". Now, if the user provides a data which matches the premise of the rule, then the truth of the conclusion will be 0.9, otherwise it will be zero.

In this paper we provide a new procedure to find a match between clauses and assign dynamically certainty factors to the conclusion. Namely, we let the matching process between clauses dictate the certainty of the conclusion.

The matching process is based on the parsing procedure described above and is divided into two parts. First, we try to match the objects in the clauses or their synonyms. If we are successful then we match the numbers associated with the objects of each clause.

This is done as follows:

Since we have to match two clauses, and since each clause may contain two distinct intervals, we actually have to assume that there may be four intervals to consider. Let the two intervals which are associated with the clause on the blackboard (or in the conclusion) be C and D, and the two possible intervals which are associated with a clause in the premise of the rule be A and B, then:

$$M = \frac{I(A,C) + I(A,D) + I(B,C) + I(B,D)}{w(C) + w(D)} \qquad (5.2)$$

where $w(\bullet) = \overline{\bullet} - \underline{\bullet} + \varepsilon$
$\overline{\bullet}$ is the upper bound of the interval,
$\underline{\bullet}$ is the lower bound of the interval, and
ε is an arbitrary small number.

The $I(\bullet)$ procedure calculates the overlapping area of any given intervals.[8] Namely,

$$I(X,Y) = \max\left(0, \min\left(\overline{X} - \underline{X} + \varepsilon, \overline{X} - \underline{Y} + \varepsilon, \overline{Y} - \underline{Y} + \varepsilon, \overline{Y} - \underline{X} + \varepsilon\right)\right) \qquad (5.3)$$

The logic behind Equation 5.3, is quite simple. We want that

$$I(X,Y) = \begin{cases} 0 & \text{if } X \text{ and } Y \text{ do not intersect} \\ \text{Equation 5.3} & \text{otherwise} \end{cases}$$

It is clear that if the two intervals do not intersect then the "min" part of the equation will generate a number which is less than zero. Maximizing that number with zero will result with a zero. If, however, the intervals do intersect, the smaller interval that is generated is the correct one. And since this interval has a positive value, this value will be retained when it is "maxed" with zero.

So the questions are "where did we get the four intervals, why did we choose the smallest one and what is the role of the e in the equation". Whenever we compare two intervals we have six cases to consider these are:

1.   •——X——•

        •——Y——•

In this case, X and Y do not intersect. The entire interval of X precedes the entire interval of Y.

2.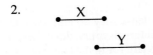

In this case the intervals do intersect and the area of the intersection is $\overline{X} - \underline{Y}$.

3.
```
  .___X___.
.___Y___.
```

Here Y is included in X so the intersection is the width of Y which is $\overline{Y} - \underline{Y}$.

4.
```
    .X.
.___Y___.
```

Here, X is included in Y so the intersection is the width of X which is $\overline{X} - \underline{X}$.

5.
```
    .___X___.
.___Y___.
```

In this case, the intersection is $\overline{Y} - \underline{X}$.

6.
```
          .___X___.
.___Y___.
```

Here, the intervals do not intersect.

Observing cases 2, 3, 4 and 5, we can see that if the two intervals do intersect then the intersection area must be smaller the any of the intervals. Thus cases 3 and 4 should be eliminated. Next we compare cases 2 and 5. If there is an intersection, then one of the two cases will generate an interval which is larger then any of the two original intervals and only the remaining case will generate the proper interval. This is the reason for choosing the smallest interval from the four available options.

Next we want to explain the reason for adding ε to the equation. Each time we compute the area of overlapping we really compute the width of an interval. Now assume that we have a single number. A single number is an interval that its lower and upper bounds are the same. In this case its width is zero. This is an error since we still have to compare two numbers. By adding the ε we are able to compute the matching factor of two number. Consider the following example:

Let the premise of some rule be "John is 20" and the conclusion of another rule be "John is 20". Since we do not have negation, then the premise clause will have only one interval (A) and the conclusion clause will have only one interval (C) such that A = [20,20] and C = [20,20]. Now, using Equation 5.3, we get:

$$
\begin{aligned}
I(A,C) &= \max(0; \min(20-20+\varepsilon; 20-20+\varepsilon; 20-20+\varepsilon; 20-20+\varepsilon)) \\
&= \max(0; \varepsilon) \\
&= \varepsilon
\end{aligned}
$$

Using Equation 5.2, we get $M = \dfrac{I(A,C)}{w(C)} = \dfrac{\varepsilon}{20-20+\varepsilon} = 1$ which is a good result.

Next we change the example. Assume the the premise clause remains the same but the conclusion clause is changed to "John is 21". Again we have only one interval for the premise (A = [20,20]) and one interval for the conclusion (C = [21,21]). Now, using Equation 5.3, we get:

$$
\begin{aligned}
I(A,C) &= \max(0; \min(20-20+\varepsilon; 21-21+\varepsilon; 20-21+\varepsilon; 21-20+\varepsilon)) \\
&= \max(0; -1+\varepsilon) \\
&= 0.
\end{aligned}
$$

Using Equation 5.2, we get $M = \dfrac{I(A,C)}{w(C)} = \dfrac{0}{21-21+\varepsilon} = 0$

Again, the proper result. Observing Equations 5.2 and 5.3, it is clear that omiting the ε will generate erroneous results.

It is also important to mention, that if we are computing the area of real intervals, the ε will not effect the computation since its value is very small $\left(\text{we choose } \varepsilon = \dfrac{1}{MAXINT}\right)$.

Using Equations 5.2 and 5.3, we create *R-List*, which is the list that contain the relations among the rules in the knowledge base. The construction of *R-List* is quite simple. We compare each conclusion clause with all the premise clauses. If

$$M(\text{conclusion clause; premise clause}) \geq T, \qquad (5.4)$$

where *T* is some threshold, then we add that relation to *R-List*.

For example, assume we have the following two clauses:

John is not more_or_less 20   John is almost 19
First we see that the objects are the same (both clauses contain the object "John"), so we

may proceed and compute the similarity between "not more_or_less 20" to "almost 19".

Using the parsing procedure describe above, we parse the two clauses and generate the proper intervals to be associated with them. Assuming that $\alpha = 0$ (this is the lower bound of the domain *OLD*) and $\beta = 100$ (this is the upper bound of the same domain). Also assume that "more_or_less" generates an interval of [X - 10% , X + 10%] and "almost" generates the interval [X - 5% , X - 1%]. Using the above definitions we get:

"not more_or_less 20" → "not [18,22]" → [0, 18 - ε] and [22 + ε , 100]
"almost 19" → [18.05 , 18.81]

Also assume that A = [0, 18 - ε], B = [22 + ε, 100] and C = [18.05, 18.81]. Thus Equation 5.2 is reduced to

$$M = \frac{I(A,C) + I(B,C)}{w(C)}$$

So,

$$\begin{aligned}
I(A,C) &= \max(0, \min(18, -0.05, 0.76 + \varepsilon, 18.81 + \varepsilon)) \\
&= \max(0, -0.05) \\
&= 0 \\
I(B,C) &= \max(0, \min(78, 81.95 + \varepsilon, 0.76 + \varepsilon, -81.19 + \varepsilon)) \\
&= \max(0, -81.19 + \varepsilon) \\
&= 0.
\end{aligned}$$

and $w(C) = 0.76 + \varepsilon$ thus $M = \dfrac{0}{0.76 + \varepsilon} = 0$

There is one exception to the procedure described above. If we have two clauses (C_1 and C_2) and
C_1 = "X is Y" where Y is a word and
C_2 = "X is N" where N is a number then
If Y (in C_1) is associated with some membership function (F) then:

$$M = x_F(N) \tag{5.5}$$

where X is the grade of membership of N in F.

Thus, if $M > T$ (where T is some threshold) then we add that relation to R - *list*.
For example, assume we have the clause "The temperature is high" where

high - temperature = $\pi(80; 120)$ in the range [50,200]

and another clause "the temperature is 100", then $x_{high\text{-}temperature}(100) = 0.88$
So, according to Equation 5.6, M = $x_{high\text{-}temperature}(100) = 0.88$
If $M > T$ then we link the two clauses and add the similarity measure to R - *list*.
Linking mathematical relations with other mathematical relation or any other English

clause is very simple. As was mentioned above, a mathematical expression in the conclusion part of any rule can be an assignment statement. Let the left hand side of the assignment statement be X. Now we have to consider two cases:

1. If the clause in the premise part of the rule is a mathematical expression then it must be a relational expression (E). In this case:

$$M = \begin{cases} -1 & \text{if } X \in E \\ 0 & \text{otherwise} \end{cases} \qquad (5.6)$$

2. If the clause in the premise part of the rule is an English proposition then:

$$M = \begin{cases} -1 \text{ if the proposition is in the form " } X \text{ is } N" \\ \quad (\text{where } N \text{ is some number}) \\ 0 \quad \text{otherwise} \end{cases} \qquad (5.7)$$

The -1 indicates that the actual matching between the two clauses will be computed during inference (since we do not know a priori the value of X in the mathematical expression).

For example, let the mathematical expression be $X := sqrt(Y)$, and the English proposition be "X is 7". It is clear that we can not match the two clause since we do not know the value of Y. During the inference, when the value of Y will be provided, the matching value between the above two clauses can be computed.

Whenever, the compiler sees an executable clause, it will not perform any matching since we can not match executable procedures. Let C be a clause then:

$$M = 0 \text{ if } C \text{ is an executable clause} \qquad (5.8)$$

5.6 CONCLUSION

In this paper we have investigated a procedure for matching any two clauses. It has been shown that the procedure to perform fuzzy matchings between two clauses is general and handles cases where the information provided by the user is fuzzy or incomplete (this information can be in the form of a rule in the knowledge base or user entered data). The generality and the simplicity of the procedure may provide a very useful tool in designing fuzzy expert systems or fuzzy expert system shells.

REFERENCES

1. Liebowitz, J. Introduction to *Expert Systems*. Mitchell Publishing, Inc. (1988).

2. Schneider, M., and A. Kandel. *Cooperative Fuzzy Expert Systems - Their Design and Applications in Intelligent Recognition.* Verlag TUV Rheinland, Koln, Germany(1988).

3. Kandel, A., and G. Langholz. *Hybrid Architectures for Intelligent Systems.* CRC (1993).

4. Schneider, M., and A. Kandel. "On Uncertainty Management in Fuzzy Inference Procedures." First International Conference on Fuzzy Theory and Technology, Durham, NC, October 14-18, 1992. (Invited Paper).

5. Gupta, M. M., A. Kandel, W. Bandler, and J. B. Kizka. *Approximate Reasoning in Expert Systems.* North-Holland (1985).

6. Negoita, C. V. *Expert System and Fuzzy Systems.* The Benjamin/Cummings Publishing Company (1985).

7. Schneider, M., E. Shnaider, and A. Kandel. *Application of the Negation Operator in Fuzzy Production Rules.* Accepted for publication in the International Magazine for Fuzzy Sets and Systems (1989).

8. Schneider, M., D. Clark, and A. Kandel. *On the Matching Process in Fuzzy Expert Systems.* 3th International IPMU Conference, Paris, France, July 2-6, 1990, pp. 46-48.

9. Kandel, A. *Fuzzy Expert Systems.* CRC (1993).

10. Good, D., and L. Ralph. "Computer Interval Arithmetic: Definition and Proof of Correct Implementation." *Jour nal of the Association for Computing Machinery.* 4 (1970): 603-612.

11. Moore, Ramon. *Mathematical Elements of Scientific Computing.* New York: Holt, Rinehart and Winston, Inc. (1975).

12. Moore, Ramon. *Methods and Applications of Interval Analysis.* Philadelphia: SIAM, (1979).

13. Kandel, A., M. Schneider, and G. Langholz. "The Use of Fuzzy Logic For the Management of Uncertainty in Intelligent Hybrid Systems." In *Uncertainty Management in Expert Systems.* L. Zadeh and J. Kacprzyk (editors), John Willey & Sons (1992), pp. 569-587.

14. Wagman, D., M. Schneider, A. Kandel, E. Shnaider, and G. Langholz. *On the Use of Logical Relations in Expert Systems.* The 9th Israeli Conference on Artificial Intelligence and Computer Vision, Tel-Aviv, Israel, December 28-29 (1992).

15. Kandel, A. and G. Langholz. *Fuzzy Control Systems.* CRC (1994).

6

COMPUTATIONAL NEURAL ARCHITECTURES FOR CONTROL APPLICATIONS

L. Jin, M.M. Gupta, and P.N. Nikiforuk
University of Saskatchewan
Saskatoon, Saskatchewan, CANADA

Advances in the area of computational neural networks have provided the potential for new approaches to the control of complex and unknown nonlinear systems through learning processes. In this chapter, some computational neural structures are introduced for the learning and control of unknown discrete-time nonlinear systems. The basic architectures and mathematical expressions of static multilayered feedforward neural networks and dynamic recurrent neural networks are discussed in detail. Some control-relevant dynamics of neural networks are explored for the purpose of designing adaptive control systems. The concept of an inverse system is incorporated in such neural control systems.

6.1 INTRODUCTION

Artificial neural networks, as models of specific biological structures, have the advantages of distributed information processing and the inherent potential for parallel computation. An artificial neural network consists of many interconnected identical simple processing units, called *neurons* or *nodes*, which form the layered configurations. An individual neuron aggregates its weighed inputs and yields an output through a nonlinear activation function with a threshold. There are three types of synaptic interconnections, *intra-layer*, *inter-layer* and *recurrent* connections, in artificial neural networks. The *intra-layer connections* also called *lateral* connections or *cross-layer* connections, are links between the neurons in the same layer of the network. The *inter-layer connections* are links between neurons in different layers. The *recurrent connections* provide self-feedback links to the neurons. In the inter-layer connections, the signals are transformed in one of two ways; either feedforward or feedback.

Advances in the area of artificial neural networks have provided the potential for new approaches to the control of complex and unknown nonlinear systems through learning processes. The main potentials of neural networks for control applications can be summarized as:

(i) they can be used to approximate any continuous mapping to any desired degree of accuracy;
(ii) they perform this approximation through the process of learning; and
(iii) parallel processing and fault tolerance are inherent characteristics of such neural networks.

The objective of neural networks based adaptive control systems for unknown nonlinear complex dynamic plants is to develop algorithms for identification and control using neural networks through a learning process. To avoid modeling difficulties for complex physical systems, neural methods of learning and control provide a natural framework for the design of tracking controllers for unknown nonlinear systems. This can be viewed as the nonlinear dynamic mapping of control inputs into observation outputs. A number of multilayered feedforward neural networks based controllers have been recently proposed (Chen, 1990, Narendra and Parthasarathy, 1990 1991, and Jin, Nikiforuk and Gupta, 1992, 1993a). For such types of adaptive learning control systems, feedforward neural networks are used to approximate the unknown nonlinear functions that are contained in the nonlinear systems, so that the adaptive control laws can then be designed on the basis of a neural approximation model.

Narendra and Parthasarathy (1990), and Hunt and Sbarbaro (1991) have proposed adaptive control approaches using neural networks for nonlinear systems without the internal dynamics. They divided the adaptive control problem for an unknown nonlinear system into an identification, or system modeling stage, and a nonlinear control stage. Jin, Nikiforuk and Gupta (1992) have provided a novel algorithm for adaptive tracking for SISO nonlinear systems using layered feedforward neural networks. They have also analyzed the convergence of the weight learning and the stability of the closed system. In fact, because of the complexities of nonlinear dynamic systems the main efforts of the neural networks based nonlinear adaptive control approaches are focused currently on SISO nonlinear systems without the internal dynamics. Other important current research topics in this field are those of neural learning and control for a general class of MIMO nonlinear systems.

More recently, it has been reported in several studies (Williams and Zipser, 1989, and Sudharsanan and Sundereshan, 1991) that an appropriate dynamic mapping may be realized by a dynamic recurrent neural network which is trained through a series-parallel or a parallel learning model, similar to the case of the feedforward networks, so that a desired response is obtained. The recurrent neural network consists of both feedforward and feedback connections between the layers and neurons forming a complicated dynamic system. Obviously, the ability of a recurrent neural network to approximate a continuous or discrete nonlinear dynamic system by neural dynamics defined by a system of nonlinear differential or difference equations has the potential for applications to adaptive control problems. When dynamic recurrent neural networks are used to approximate and control an unknown nonlinear system through on-line learning processes, they may be treated as subsystems of such adaptive control systems, where the weights of the networks need to be updated using a dynamic learning algorithm during the control processes.

In this chapter, some computational neural structures are introduced for purposes of learning and controlling unknown discrete-time nonlinear systems. In Section 6.2, the basic architectures and mathematical expressions of *multilayered feedforward neural networks* (MFNNs) are discussed. The *tapped delay neural networks* (TDNNs)'' which consist of MFNNs and the tapped delay lines for the network input and output are presented in Section 6.3. As well, adaptive inverse control approaches for a class of input-output nonlinear systems are studied using the TDNNs. A novel *multilayered recurrent neural network* (MRNN) structure is introduced in Section 6.4, and the input-output dynamic equations of the MRNN are derived. Some control-relevant dynamics of a general class of recurrent neural networks (RNNs) are explored in Section 6.5, and the inverse system approach for discrete-time nonlinear systems which are described using state-space equations is expressed. Dynamic learning and adaptive control techniques for unknown nonlinear systems are developed using RNNs in Section 6.6.

6.2 MULTILAYERED FEEDFORWARD NEURAL NETWORKS (MFNN)

6.2.1 Architecture and Operational Equations

As a computational architecture, the static multilayered feedforward neural network (MFNN) structure, which is one of the main classes of artificial neural networks, plays an important role in problems such as system identification, control, channel equalization, and pattern recognition. The backpropagation (BP) algorithm (Rumelhart and McCelland, 1986, and Hecht-Nielsen, 1989) is based on the gradient descent algorithm and has been used widely to train multilayered neural networks for performing desired tasks. As well, several extensions of this algorithm are also described. The advantages of BP algorithm include parallel computation, the ability to store many more patterns than the number of network inputs, and the ability to acquire arbitrarily complex nonlinear mappings.

In a multilayered feedforward network the neurons are organized into layers with non feedback or cross connections. A basic structure of multilayered feedforward neural networks (MFNNs) with feedforward connections is shown in Figures 6.1 and 6.2. Let M be the total number of layers of the MFNN including the input and output layers. Let n_s be the total number of neurons in the s - th layer, and the i - th neuron in the s - th layer be denoted by neuron (s, i). The scheduling of the network's operation starts by inserting the input vector u into the first layer, called the input layer. The processing elements of the first layer transmit all of the components of u to all of the units of the second layer of the network.

The outputs of the second layer are then transmitted to all of the units of the third layer, until finally the n_M outputs $x = y_i$ are formed.

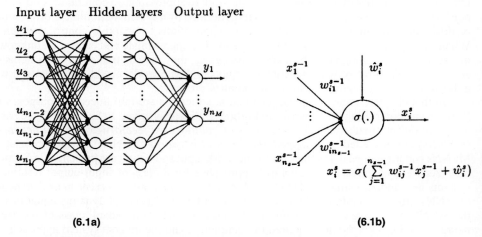

(6.1a)   (6.1b)

Figure 6.1. The multilayered feedforward neural network (MFNN): (6.1a) The network structure; (6.1b) The *i - th* neuron of the *s - th* layer.

Mathematically, the operation of the neuron (s, i) is defined as

$$z = \begin{cases} u_i & \text{if } s = 1 \\ w_{i,k}^{s-1} x_k^{s-1}, & \text{if } 2 \leq s \leq M \end{cases} \quad (6.1)$$

and

$$x = \begin{cases} \sum_{k=1}^{n_{s-1}} z_{i,k}^s, & \text{if } s = 1 \text{ or } M \\ \sigma(\sum_{k=1}^{n_{s-1}} z_{i,k}^s + \hat{w}_i^s), & \text{if } 2 \leq s \leq M - 1 \end{cases} \quad (6.2)$$

where u_i is the input of the neuron $(1, i)$, x is the output of the neuron (s, i), w is the linkweight coefficient from the *neuron* (s, k) to the *neuron* $(s + 1, i)$, is the threshold of the *neuron* (s, i) and $\sigma(\cdot)$ is the neuron's nonlinear activation function. As a biological equivalence, Eqn. (6.1) represents a synaptic operation and Eqn. (6.2) represents a somatic operations as shown in Figure 6.2. Indeed, it is easily seen that a multidimensional output vector from the layer $(i - 1)$ is directly used as an input vector to the layer i, assuming that a neuron is capable of processing a vector signal. This fact is also shown in Figure 6.2.

Let y_i, $1 \leq i \leq n_M$ be the output of the network. The output equations may then be represented as

$$y_i = x_i^M = \sum_{k=1}^{n_{M-1}} w_{i,k}^{M-1} x_k^{M-1}, \quad 1 \leq i \leq n_M \qquad (6.3)$$

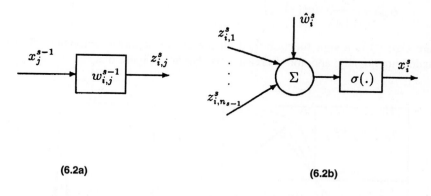

      (6.2a)   (6.2b)

Figure 6.2. Mathematical operation of the *neuron (s, i)*:(6.2a) The synaptic operation; (6.2b) The somatic operation.

6.2.2 Nonlinear Input-Output Mapping

The neural activation function $\sigma(\cdot)$ in the somatic operation may be chosen as a continuous and differentiable nonlinear sigmoidal function satisfying the following conditions:

(i) $\sigma(x) \to \pm 1$ as $x \to \pm \infty$;
(ii) $\sigma(x)$ is bounded with the upper bound 1 and the lower bound -1;
(iii) $\sigma(x) = 0$ at a unique point $x = 0$;
(iv) $\sigma'(x) > 0$ and $\sigma'(x) \to 0$ as $x \to \pm \infty$;
(v) $\sigma'(x)$ has a global maximal value $c \leq 1$.

Typical examples of such a neural activation function $\sigma(\cdot)$ are

$$\sigma(x) = \frac{e^x - e^{-x}}{e^x + e^{-x}} = \tanh(x)$$

$$\sigma(x) = \frac{1 - e^{-x}}{1 + e^{-x}}$$

$$\sigma(x) = \frac{2}{\pi} \tan^{-1}\left(\frac{\pi}{2} x\right)$$

$$\sigma(x) = \frac{x^2}{1 + x^2} sign(x)$$

where sign (·) is a sign function, and all the above nonlinear activation functions are bounded, monotonic, and non-decreasing as shown in Figure 6.3.

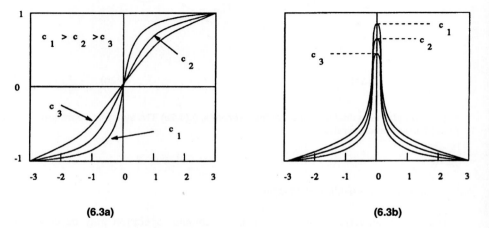

Figure 6.3. Sigmoid function σ(c x) and its derivative σ'(c x): (6.3a) $\sigma(c_1 x)$, $\sigma(c_2 x)$ and $\sigma(c_3 x)$, where $c_1 > c_2 > c_3$; (6.3b) $\sigma'(c_1 x)$, $\sigma'(c_2 x)$ and $\sigma'(c_3 x)$.

Let the neural input and output vectors be represented as

$$u = \begin{pmatrix} u_1 \\ \vdots \\ u_{n_I} \end{pmatrix} \in R^{n_I}, \quad y = \begin{pmatrix} y_1 \\ \vdots \\ y_{n_M} \end{pmatrix} \in R^{n_M}$$

Furthermore, let the augmented output vector of the s - th layer and the augmented weight matrix be represented as

$$x_a^s = \begin{pmatrix} x^s \\ \vdots \\ 1 \end{pmatrix}, \quad w_{a,i}^{s-1} = \begin{pmatrix} w_{i,1}^{s-1} \\ \vdots \\ w_{i,n_{s-1}+1}^{s-1} \end{pmatrix}, \quad W_a^{s-1} = \begin{pmatrix} \left(w_{a,1}^{s-1}\right)^T \\ \vdots \\ \left(w_{a,n_s}^{s-1}\right)^T \end{pmatrix}$$

where x_a^s is a $(n_s + 1) \times 1$ vector, and W_a^{s-1} is a $(n_s) \times (n_{s-1} + 1)$ matrix. The introduction of the augmented output vectors and the augmented weight matrices is due to the existence of the thresholds in the nonlinear neural activation function.

The input-output relationship of a MFNN may then be represented using the nonlinear mapping as follows

$$y = W_a^{M-1} \sigma_a \left(W_a^{M-2} \sigma_a \left(\ldots W_a^2 \sigma_a \left(W_a^1 x_a^1 \right) \ldots \right) \right)$$

$$= W^{M-1} \sigma \left(W_a^{M-2} \sigma_a \left(\ldots W_a^2 \sigma_a \left(W_a^1 x_a^1 \right) \ldots \right) \right)$$

$$\equiv F\left(W_a^1, W_a^2, \ldots, W^{M-1}, u \right) \tag{6.4}$$

where $x_a^1 = \left(u^T : 1 \right)^T$ and the augmented vector function

$$\sigma_a(\cdot) = \begin{pmatrix} \sigma(\cdot) \\ \vdots \\ 1 \end{pmatrix}$$

The function $F(\cdot)$ in Eqn. (6.4) is a continuous and differentiable nonlinear mapping from the input space to the output space since the nonlinear activation function $\sigma_a(\cdot)$ is continuous and differentiable. Therefore, the nonlinear mapping $F(\cdot)$ in Eqn. (6.4), which contains many synaptic weights, may be considered as a nonlinear neural mapping function from the input pattern space to the output pattern space where the mapping is done through the process of learning as opposed to being pre-programmed as done in conventional methods. In other words, in the MFNN architecture, the input information is fed forward recursively to the higher hidden layers, and finally to the output layer. It is for this reason that these networks are also called the *propagation networks* (Hecht-Nielsen, 1989). Since the input-output relationship of a MFNN is described by some static algebraic manipulations, the MFNNs are static neural networks without the internal dynamics. However, as discussed in the later sections, they may be employed in a control system as some part of a dynamic system.

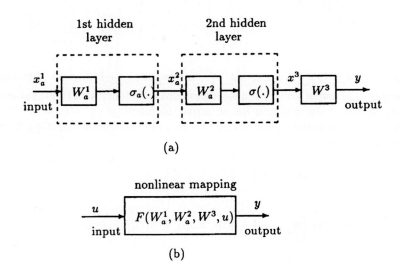

Figure 6.4. Topologic structure of a four-layered feedforward neural network with one input layer, two hidden layers, and one output layer, where the vector operational equation is $y = W^3 \sigma\left(W_a^2 \sigma_a\left(W_a^1 x_a^1\right)\right) \equiv F\left(W_a^1, W_a^2, W^3, u\right)$.

The operational equation of a commonly used three-layered feedforward neural network with a n_1 - dimensional input $u = \left(u_1, \ldots, u_{n_1}\right)^T$ and a n_3 - dimensional output $y = \left(y_1, \ldots, y_{n_3}\right)^T$ as shown in Figure 6.5a is given by

$$y_j = \sum_{i_2=1}^{n_2} w_{j,i_2}^2 \sigma\left(\sum_{i_1=1}^{n_1} w_{i_2,i_1}^1 u_{i_1} + \hat{w}_{i_2}^2\right), \quad j = 1, 2, \ldots, n_3$$

where n_2 is the number of the hidden neurons. Moreover, the above neural equation can be represented easily by the following simple vector form

$$y = W^2 \sigma\left(W_a^1 x_a^1\right) \tag{6.5}$$

where $x_a^1 = \left(u^T : 1\right)^T$ is a $(n_1 + 1)$-dimensional augmented input vector. The weight matrices of the hidden layer and the output layer are given by

$$W_a^1 = \begin{pmatrix} w_{1,1}^1 & \cdots & w_{1,n_2}^1 & \hat{w}_1^2 \\ \vdots & \vdots & \vdots & \vdots \\ w_{n_3,1}^1 & \cdots & w_{n_3,n_{23}}^1 & \hat{w}_{n_2}^2 \end{pmatrix},$$

and

$$W^2 = \begin{pmatrix} w_{1,1}^2 & \cdots & w_{1,n_2}^2 \\ \vdots & \vdots & \vdots \\ w_{n_3,1}^2 & \cdots & w_{n_3,n_2}^2 \end{pmatrix}$$

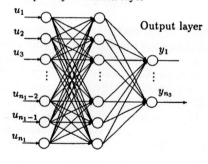

(6.5a)

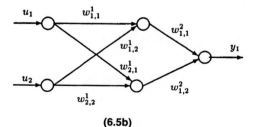

(6.5b)

Figure 6.5. The three-layered feedforward neural network: (6.5a) The structure of the network; (6.5b) The weights of the network.

6.3 TAPPED DELAY NEURAL NETWORKS FOR CONTROL APPLICATIONS

6.3.1 Neural Network Structure

Some typical structures of *tapped delay neural networks* (TDNNs) which consist of MFNNs and some time delay operators with feedback are shown in Figures 6.6 and 6.7. For time series analysis, the one-step and q-step prediction equations of the TDNNs can be given as

$$y(k+1) = F(w, y(k), ..., y(k-n), u(k)) \qquad (6.6)$$

and

$$y(k+q) = F(w, y(k), ..., y(k-n), u(k)) \qquad (6.7)$$

respectively, where $F(\cdot)$ is a continuously and differentiable function which may be obtained from the operation equations of the MFNN given in the last section, and the input components of the neural networks are the time delayed versions of the outputs of the networks or the time series systems. In this case, Eqns. (6.6) and (6.7) represent a one-step ahead nonlinear predictor and a q-step ahead nonlinear predictor, respectively.

For identification and control applications, the input-output equations of TDNNs with relative degree one and q as illustrated in Figure 6.7 are obtained as

$$y(k+1) = F(w, y(k), ..., y(k-n), u(k), ..., u(k-m)) \qquad (6.8)$$

and

$$y(k+q) = F(w, y(k), ..., y(k-n), u(k), ..., u(k-m)) \qquad (6.9)$$

respectively, where the inputs to the networks are the time delayed terms of the neural outputs and the current neural inputs. These neural network structure have the potential to represent a class of nonlinear input-output mapping of unknown nonlinear systems without internal dynamics, and have been successfully applied to the design of adaptive control systems (*Narendra* and *Parthasarathy*, 1990). Because there are no state feedback connections in the network, the static back propagation (BP) learning algorithm may be used to train the TDNN so that the processes of system modeling or function approximation are carried out, even though the TDNN is a type of dynamic neural network.

An important issue of the TDNNs is the stability of the equilibrium points in the sense of Lynapunov. Without loss of generality, it is seen that Eqn. (6.9) represents a general form of TDNNs. A state space expression of Eqn. (6.9) can be constructed from observed quantities by using the method of delay coordinates. For this, let

$$\begin{cases} x_1^a(k) = y(k) \\ \quad \vdots \\ x_{n+1}^a(k) = y(k-n) \\ x_1^b(k) = u(k-1) \\ \quad \vdots \\ x_m^b(k) = u(k-m) \end{cases}$$

Computational Neural Architectures for Control Applications

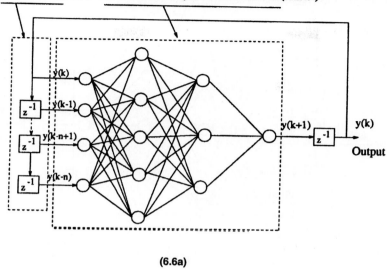

(6.6a)

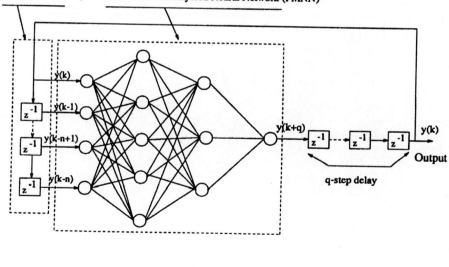

(6.6b)

Figure 6.6. Tapped delay neural networks TDNNS for time series analysis; (6.6a) TDNN with one-step prediction; (6.6b) TDNN with q-step prediction.

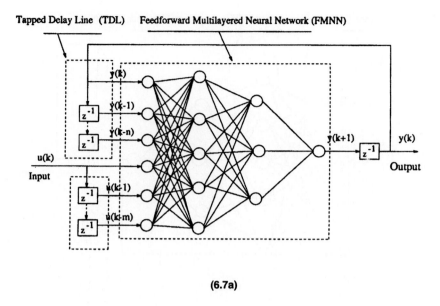

(6.7a)

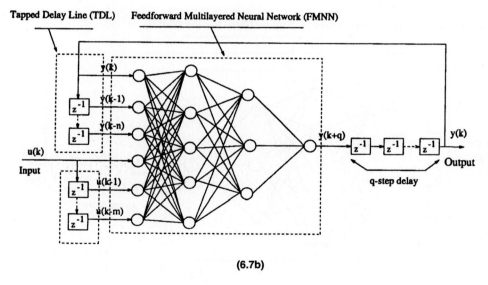

(6.7b)

Figure 6.7. Tapped delay neural networks (TDNNs) for identification and control; (6.7a) TDNN with relative degree one; (6.7b) TDNN with relative degree q.

and

$$x^a = [x_1^a, \ldots, x_{n+1}^a]^T, \quad x^b = [x_1^b, \ldots, x_m^b]^T, \quad x = \begin{bmatrix} x^a \\ x^b \end{bmatrix}$$

Thus,

$$\begin{cases} x(k+1) = \phi[x(k), u(k)] \\ y(k) = cx(k) \end{cases} \quad (6.10)$$

where

$$\phi[x(k), u(k)] = \begin{bmatrix} F[x(k), u(k)] \\ \vdots \\ x_n^a(k) \\ u(k) \\ \vdots \\ x_{m-1}^b(k) \end{bmatrix}$$

$$c = [1 0 \cdots 0 \vdots 0 \cdots 0]$$

If the external input $u(k)$ is designed as a state feedback $u(k) = u[x(k)]$ such that the neural system response converges to the equilibrium state $x^* = \phi[x^*, u(x^*)]$, then the local stability of the equilibrium point x^* can be determined using the following Jacobian

$$J(x*) = \begin{bmatrix} \alpha_1 & \cdots & \alpha_{n+1} & \beta_1 & \cdots & \beta_m \\ & & 0 & 0 & & 0 \\ & I_n & \vdots & & 0 & \vdots \\ & & 0 & & & 0 \\ v_1 & \cdots & v_{n+1} & \mu_1 & \cdots & \mu_m \\ & & 0 & & I_{m-1} & \vdots \\ & & & & & 0 \end{bmatrix}$$

where

$$\alpha_i = \left[\frac{\partial F(x,u)}{\partial x_i^a} + \frac{\partial F(x,u)}{\partial x_u}v_i\right]\Bigg|_{x=x^*, u=u(x^*)}$$

$$\beta_i = \left[\frac{\partial F(x,u)}{\partial x_i^b} + \frac{\partial F(x,u)}{\partial x_u}\mu_i\right]\Bigg|_{x=x^*, u=u(x^*)}$$

$$v_i = \frac{\partial u}{\partial x_i^a}\Bigg|_{x=x^*}$$

$$\mu_i = \frac{\partial u}{\partial x_i^b}\Bigg|_{x=x^*}$$

An equilibrium state x^* of system (6.10) is locally asymptotically stable in the sense of Lyapunov if the eigenvalues of $J(x^*)$ are located inside the unit circle. On the other hand, if the external input $u(k)$ is a constant, $u(k) = u^*$, $k = 0, 1,...$, the Jacobian $J(x^*)$ becomes

$$J(x^*) = \begin{bmatrix} \alpha_1 & \cdots & \alpha_{n+1} & \beta_1 & \cdots & \beta_m \\ & & 0 & 0 & & 0 \\ & I_n & \vdots & & 0 & \vdots \\ & & 0 & & & 0 \\ 0 & \cdots & 0 & 0 & \cdots & 0 \\ & & 0 & I_{m-1} & & \vdots \\ & & & & & 0 \end{bmatrix}$$

Hernandez and Arkun (1992) showed that the eigenvalues of $J(x^*)$ consist of zeros with multiplicity m and the eigenvalues of the following matrix

$$J_y(x^*) = \begin{bmatrix} \alpha_1 & \cdots & \alpha_{n+1} \\ & & 0 \\ & I_n & \vdots \\ & & 0 \end{bmatrix}$$

where $J_y(x^*)$ is the Jacobian of $x^a(k+1)$ with respect to $x^a(k)$. In this case, an equilibrium point x^* of the system (6.10) is locally asymptotically stable if the eigenvalues of $J_y(x^*)$ are contained inside the unit circle.

6.3.2 Adaptive Control Using TDNNs

Consider a SISO unknown nonlinear system which has the following input-output equation expressed in a canonical form

$$y_p(k+1) = f(y_p(k), ..., y_p(k-n), u(k), ..., u(k-m))$$

$$= f(x(k), u(k)) \quad (6.11)$$

where $x(k) = [y_p(k), ..., y_p(k-n), u(k-1), ..., y(k-m)]^T$ is a state vector, and $f(\cdot)$ is a unknown nonlinear function that satisfies $\partial f(x, u)/\partial u \neq 0$. The *canonical form* of Eqn. (6.11) represents a general class of input-output nonlinear systems without the *internal dynamics*. For a desired output $y_d(k)$, some control schemes for the purpose of adaptive tracking using TDNNs will now be discussed in this subsection.

In direct inverse control (DIC), as shown in Figure 6.8, the input-output equation of the TDNN which produces the control signal to the system is expressed as

$$u(k) = F(w, x(k), r(k+1)) \quad (6.12)$$

where $r(k+1)$ is a reference input defined by

$$r(k+1) = y_d(k+1) + \sum_{i=1}^{\gamma} \beta_i (y_d(k-i+1) - y_p(k-i+1)), \quad \gamma \leq k \quad (6.13)$$

Let the nonlinear mapping $F(\cdot)$ be trained to approximate the inverse system of the unknown nonlinear system; that is

$$F(w, x, r) \rightarrow f_u^{-1}(x, r)$$

where $f_u^{-1}(x, r)$ satisfies

$$y_p(k+1) = f(x(k), u(k))$$

$$= f(x(k), f_u^{-1}(x(K), r(k+1)))$$

$$= r(k+1) \quad (6.14)$$

Then

$$y_p(k+1) = f(x(k), u(k)) = f(x(k), F(w, x(k), r(k+1))) \rightarrow r(k+1)$$

The output tracking control can be accomplished as long as the parameters β_i, $i = 1, ..., \gamma$ are chosen so that the roots of the characteristic equation

$$z^\gamma + \beta_\gamma z^{\gamma-1} + ... + \beta_1 = 0$$

lie inside the unit circle. If all $\beta_i = 0$, $i = 1, 2, ..., \gamma$ are chosen, then Eqn. (6.14) becomes

$$y_p(k+1) \rightarrow y_d(k+1)$$

which describes an output dead-beat response. In terms of robustness, the dead-beat response is not a good choice although the closed-loop response of such a system is the fastest achievable closed-loop response in terms of tracking the desired output $y_d(k)$ (Isermann, 1989). The error index used to train the TDNN is defined as

$$E(k) = \tfrac{1}{2}(r(k) - y_p(k))^2 = \tfrac{1}{2}e^2(k) \qquad (6.15)$$

and the partial derivative of $E(k)$ with respect to the weight vector is obtained as

$$\frac{\partial E}{\partial w} = -e\frac{\partial y_p}{\partial F}\frac{\partial F}{\partial w} \qquad (6.16)$$

If the sign of the $\partial y_p / \partial u$ is known, the BP learning algorithm may be applied to train the TDNN so that the output tracking control is achieved.

Although its simpler structure and need for less computation are significant advantages of the direct inverse control (DIC) scheme, more a prior knowledge about the unknown plant is needed.

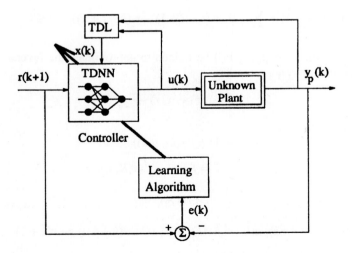

Figure 6.8. Direct inverse control (DIC) using TDNN.

The indirect inverse control (IIC) scheme is shown in Figure 6.9. The input-output equation of the TDNN is

$$y_n(k+1) = F(x(k), u(k)) \quad (6.17)$$

Let this TDNN be used to approximate the unknown plant through a weight learning process; that is

$$F(w, x, y) \to f(x, u)$$

The controller may then be obtained by inverting the TDNN equations as follows

$$u(k) = F_u^{-1}(w, x(k), r(k+1)) \quad (6.18)$$

which, implicitly, is the inverse of the input-output equation of the TDNN with respect to current control $u(k)$. Let the reference input $r(k+1)$ be designed as

$$r(k+1) = y_d(k+1) + \sum_{i=1}^{\gamma} \beta_i (y_d(k-i+1) - y_n(k-i+1)), \quad \gamma \le k \quad (6.19)$$

where the parameters β_i, $i = 1, ..., \gamma$ are chosen so that the roots of the characteristic equation

$$z^\gamma + \beta_\gamma z^{\gamma-1} + ... + \beta_1 = 0$$

lie inside the unit circle. Substituting Eqn. (6.18) into Eqn. (6.17) yields

$$\begin{aligned} y_n(k+1) &= F(w, x(k), u(k)) \\ &= F(w, x(k), F_u^{-1}(w, x(k), r(k+1))) \\ &= r(k+1) \\ &= y_d(k+1) + \sum_{i=1}^{\gamma} \beta_i (y_d(k-i+1) - y_n(k-i+1)), \gamma \le k \end{aligned} \quad (6.20)$$

that is

$$\lim_{k \to \infty} (y_d(k) - y_n(k)) = 0$$

On the other hand, if the output $y_p(k)$ of the unknown plant is perfectly approximated by the output $y_n(k)$ of the TDNN through a learning process

$$\lim_{k \to \infty} (y_p(k) - y_n(k)) = 0$$

The output tracking error of the unknown plant with respect to the desired output then satisfies

$$| y_d(k) - y_p(k) | \leq y_d(k) - y_n(k) | + | y_p(k) - y_n(k) | \to 0$$

In this case, the error index is represented as

$$\begin{aligned} E(k) &= \tfrac{1}{2}(r(k) - y_p(k))^2 \\ &= \tfrac{1}{2}(y_n(k) - y_p(k))^2 \\ &= \tfrac{1}{2}(F(w, x(k-1), u(k-1)) - y_p(k))^2 \end{aligned} \quad (6.22)$$

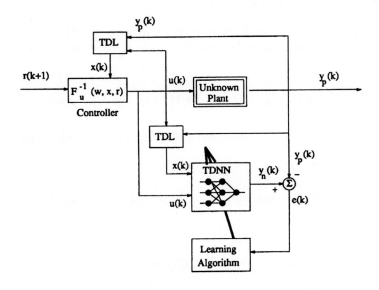

Figure 6.9. Indirect inverse control (IIC) using TDNN.

6.4 MULTILAYERED RECURRENT NEURAL NETWORKS

From a computational point of view, a dynamic neural structure which contains a state feedback may provide more computational advantage than that of a purely feedforward neural structure. For some problems, a small feedback system is equivalent to a large and possible infinite feedforward system (Hush and Horne, 1993). A nonlinear dynamic recurrent neural network structure is particularly appropriate for identification, control and filtering applications. In this section, a novel multilayered recurrent neural network (MRNN) architecture is proposed. The MRNN is a hybrid feedforward and feedback neural network, with the feedback represented by the recurrent connections and cross-talk, appropriate for approximating a nonlinear dynamic system. The MRNN is composed of an input layer, a series of hidden layer, and an output layer. It allows for feedforward and feedback among the neurons of neighboring layers, and cross-talk and recurrency in the hidden layers.

A basic structure of multilayered recurrent neural networks (MRNNs) with feedforward and feedback connections is shown in Figure 6.10. Let M be the total number of hidden layers of the MRNN, the i - th neuron in the s - th hidden layer be denoted by *neuron* (s, i), N_s be the total number of neurons in the s - th hidden layer, u_i be the i - th output of the MRNN, $x_{s,i}(k)$ be the state of the *neuron* (s, i), y_i be the i - th output of the MRNN, w be the intra-layer linkweight coefficient from the *neuron* (s, j) to the *neuron* (s, i), w be the feedforward linkweight coefficient from the *neuron* $(s - 1, j)$ to the *neuron* (s, i), w be the feedback linkweight coefficient from the *neuron* $(s + 1, j)$ to the *neuron* (s, i), and w be the threshold of the *neuron* (s, i). Mathematically, the operation of the *neuron* (s, i) is defined by the following dynamic equations.

For the first hidden layer:

$$x_{1,i}(k+1) = \sigma\left[\sum_{i=1}^{N_2} w_{2,j}^{1,i} x_{2,j}(k) + \sum_{j=1}^{N_1} w_{1,j}^{1,i} x_{1,j}(k) + \sum_{i=1}^{I} w_{0,j}^{1,i} u_j(k) + w_T^{1,i}\right]$$

$$i = 1, 2,..., N_1 \tag{6.23}$$

For the s - th hidden layer:

$$x_{s,i}(k+1) = \sigma\left[\sum_{i=1}^{N_{s+1}} w_{s+1,j}^{s,i} x_{s+1,j}(k) + \sum_{j=1}^{N_s} w_{s,j}^{s,i} x_{s,j}(k) + \sum_{i=1}^{N_{s-1}} w_{s-1,j}^{s,i} x_{s-1,j}(k) + w_T^{1,i}\right]$$

$$= \sigma\left[\sum_{h=-1}^{1} \sum_{j=1}^{N_{s+h}} w_{s+hj}^{s,i} x_{s+hj}(k) + w_T^{s,i}\right]$$

$$s = 2, 3,..., M; \quad i = 1, 2,..., N_s \tag{6.24}$$

There are no feedback actions from the output layer in the M - th hidden layer; that is, $w_{M+1,j}^{M,i} \equiv 0$. If the activation function $\sigma(\cdot)$ is a *symmetric ramp function*, the MRNN is then a special type of the brain-state-in-a-box (BSB) model with a nonsymmetric weight matrix. The terms on the right-hand side of the above equation represent respectively the feedback from the upper hidden layer, the intra-layer connections, and the feedforward from the lower layer. The output equations of the MRNN are given by

$$y_i(k) = \sum_{j=1}^{N_M} w_{M,j}^{M+1,i} x_{M,j}(k), \quad i = 1,2,...,m \tag{6.25}$$

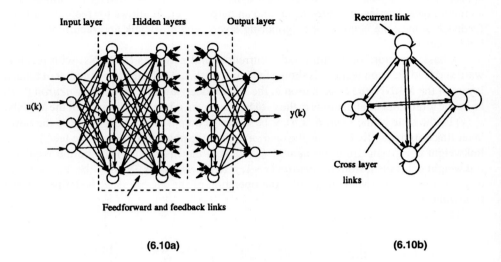

Figure 6.10. Multilayered recurrent neural network (MRNN): (6.10a) Network structure; (6.10b) The intra-layer connections.

Since the function $-1 \leq \sigma(\cdot) \leq 1$, the state vector $x(k)$ of the system (6.23-6.25) exists in a n-dimensional hypercube $H^n = [-1, 1]^n$, and the output $y(k)$ is uniformly bounded for the bounded input $u(k)$. For the adaptive learning control purposes, the number of input u_i of MRNN is assumed to have the same number as the output y_i. Furthermore, in order to obtain the I/O relationship of the MRNN shown in Figure 6.10, let $x_i = [x_{i,1}\ x_{i,2}\ ...,\ x_{i,ni}]^T$ be the state of the neurons in the i-th hidden layer of the MRNN. The dynamic neural system (6.23-6.25) can then be represented in the form of vector difference equations.

$$x_1(k+1) = \sigma\left[W_1^1 x_1(k) + W_2^1 x_2(k) + W_0^1 u_1(k) + W_T^1\right]$$

$$\equiv \sigma_1[x_1(k), x_2(k), u(k)]$$

$$x_i(k+1) = \sigma\left[W_{i-1}^i x_{i-1}(k) + W_i^i x_i(k) + W_{i+1}^i x_{i+1}(k) + W_T^i\right]$$

$$\equiv \sigma_i[x_{i-1}(k), x_i(k) + x_{i+1}(k)], \quad i = 1, 2, ..., M-1$$

$$x_M(k+1) = \sigma\left[W_{M-1}^M x_{M-1}(k) + W_M^M x_M(k) + W_T^M\right]$$

$$\equiv \sigma_M[x_{M-1}(k), x_M(k)]$$

$$y(k) = W_{M+1}^{M+1} x_M(k)$$

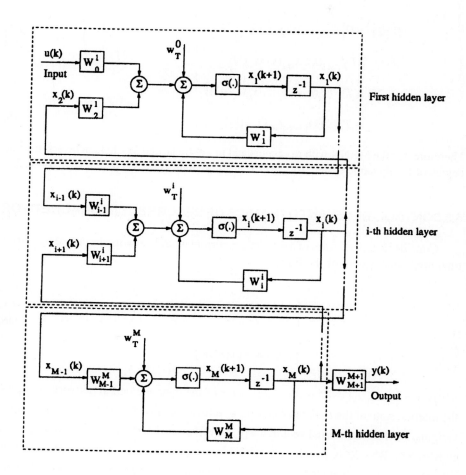

Figure 6.11. The block diagram of multilayer recurrent neural network (MRNN).

The block diagram of the MRNN is given in Figure 6.11. Moreover, the relationship between the state $x(k)$, input $u(k)$ and output $y(k)$ may be derived as follows:

$$y(k+1) = W_{M-1}^{M+1}\sigma_M\big[x_{M-1}(k), x_M(k)\big]$$

$$\equiv T^1\big[x_{M-1}(k), x_M(k)\big],$$

$$y(k+2) = T^1\big[x_{M-2}(k), x_{M-1}(k), x_M(k)\big]$$

$$\equiv T^2[x_{M-2}(k), x_{M-1}(k), x_M(k)]$$

$$y(k+M) = T^{M-1}[x_1(k+1), ..., x_M(k+1)]$$

$$\equiv T^M[x_1(k), ..., x_M(k), u(k)] \tag{6.26}$$

Therefore, for the MRNN with m - inputs and m - outputs, and M - hidden layers, the relative degree of the dynamic system (6.23-6.25) at some point (x^0, u^0) is then $r^j = M$, $j = 1, 2,..., m$.

6.5 CONTROL-RELEVANT DYNAMICS OF RECURRENT NEURAL NETWORKS

Consider a general form of a dynamic recurrent neural network described by a discrete-time nonlinear system of the form

$$\begin{cases} x(k+1) = f(x(k), u(k), w) \\ y(k) = h(x(k), w) \end{cases} \tag{6.27}$$

where $x = [x_1, ..., x_n] \in R^n$ is the state vector of the dynamic neural network, x_i represents the internal state of the i-th neuron, w is $\bar{n}$-dimensional vector of the synaptic connection weights, $u \in R^m$ is a external vector, $y = [y_1,..., y_m]^T$ is an observation vector or output vector, f: $R^n \times R^m \times R^{\bar{n}} \to R^n$ is a continuous and differentiable vector valued function, $f_i(\cdot)$ and $\partial f_i / \partial x$ are respectively bounded and uniformly bounded, and h(x): $R^n \times R^{\bar{n}} \to R^m$ is a known continuous and differentiable vector valued function.

Let $h_j(f(x, u, w))$ be independent of u; that is, $h_j(f(x, u, w)) = (h_j \circ f)(x, w)$. Then the derivative $\partial[h_j(f(x, u, w))] / \partial u = 0$ for all (x, u) in an open and dense subset O_i of $S \times U$. One recursively defines $h_j \circ f^i(x, u, w) := (h_j \circ f^{i-1})[f(x, u, w)]$, $i = 1, 2, ..., r_j$, where, in general, one may assume that $h_j \circ f, h_j \circ f^2, ...$ etc., are independent of u, up to r_j compositions. This leads to the following concept of *relative degree*. A MIMO system of the form (6.27) with the weight vector w has a (vector) relative degree $\{r_1, r_2, \ldots r_m\}$ at a point (x^0, u^0) if (i). $\partial[h_j \circ f^k(x, u, w)] / \partial u = 0$; for all $1 \leq j \leq m$ and $k < r_j - 1$, and for all x in a neighborhood of (x^0, u^0). $\partial[h_j \circ f^{r_j}(x, u, w)] / \partial u \neq 0$ for at least one $1 \leq j \leq m$.

The relative degree of the system is feedback invariant; that is, it can not be modified by feedback (Isidori, 1989, and Nijmeijer, 1990). Indeed, the relative degree r_j determines the time delay undergone by the input signals u before they influence the output y_j of the system. If system (6.27) is of relative degree $\{r_1, \ldots, r_m\}$, then $r = r_1 + \ldots + r_m \leq n$ (Isidori, 1989 and Nijmeijer, 1990).

Set

$$\begin{cases} \phi_1^j(x, w) = h_j(x, w) \\ \phi_2^j(x, w) = h_j \circ f(x, w) \\ \phi_{r_j}^j(x, w) = h_j \circ f^{r_j-1}(x, w) \\ j = 1, \ldots, m \end{cases} \qquad (6.28)$$

If $r = r_1 + \ldots + r_m$ is strictly less than n, it is always possible to find other $n - r$ functions $\phi_{r+1}(x, w), \ldots, \phi_n(x, w)$ such that the mapping

$$\Phi(x, w) = \begin{bmatrix} \phi_1^1(x, w) \\ \vdots \\ \phi_{r_m}^m(x, w) \\ \phi_{r+1}(x, w) \\ \vdots \\ \phi_n(x, w) \end{bmatrix} \qquad (6.29)$$

has a Jacobian matrix which is nonsingular at x^0. In other words, the functions $\phi_{r+1}(x, w)$, $\ldots, \phi_n(x, w)$ are arbitrary functions chosen to be independent of the functions $\phi_1^j(x, w)$, $\phi_2^j(x, w), \ldots, \phi_{r_j}^j(x, w), j = 1, 2, \ldots m$. Therefore, for the weight vector w, Φ is a diffeomorphism on a neighborhood of x^0. Setting $z(k) = \Phi(x(k), w)$, the system (6.27) can be transformed into a *normal form* in the new coordinate $z(k)$ as follows:

$$\begin{cases} z_1^j(k+1) = z_2^j, \quad j = 1, 2, ..., m \\ z_2^j(k+1) = z_3^j \\ \vdots \\ z_{r_j}^j(k+1) = \left[h_j \circ f^{r_j}\right](\Phi^{-1}(z(k)), u(k), w) \\ z_{r+1}(k+1) = q_1(u(k), z(k), w) \\ \vdots \\ z_n(k+1) = q_{n-r}(u(k), z(k), w) \\ y_j(k) = z_1^j(k) \end{cases} \quad (6.30)$$

where $q_i(z) = \phi_i(\Phi^{-1}(z, w))$ for all $r + 1 \leq i \leq n$.

Let $\xi = [z_1^1, ..., z_{r_m}^1, ..., z_1^m, ..., z_{r_m}^m]^T$, $\eta = [z_{zr+1}, ..., z_n]^T$, and the system (6.27) have the initial condition $x(0) \in h^{-1}(0) = \{x \in U: h(x, w) = 0\}$, where $h^{-1}(0)$ is a sufficiently smooth sliding manifold be controlled by feedback control $u(k) = \alpha(z(k), w)$ such that

$$z_{r_j}^j(k+1) = 0$$

that is,

$$\left[h_j \circ f^{r_j}\right](\Phi^{-1}(z(k)), \alpha(z(k), w), w) = 0 \quad (6.31)$$

for all k. The subsystem

$$\eta(k+1) = q(0, \eta(k), w) \quad (6.32)$$

is then addressed as having *"zero dynamics"*. If this subsystem is asymptotically stable, system (6.27) is said to be *minimum phase* (Sira-Ramirez, 1991).

The relations between the future outputs and current state may be derived as

$$\begin{cases} y_1(k+i_1) = h_1 \circ f^{i_1-1}(x(k), w) \; i_1 = 1, ..., r_1 \\ \vdots \\ y_m(k+i_m-1) = h_m \circ f^{i_m-1}(x(k), w) \; i_m = 1, ..., r_m \end{cases} \quad (6.33)$$

On the other hand, the equation between output and input is obtained as

$$\begin{cases} y_1(k+r_1) = h_1 \circ f^{r_1}(x(k), u(k), w) \\ \qquad\quad \equiv g_1(x(k), u(k), w) \\ \qquad\quad \vdots \\ y_m(k+r_m) = h_m \circ f^{r_m}(x(k), u(k), w) \\ \qquad\quad \equiv g_m(x(k), u(k), u(k), w) \end{cases} \quad (6.34)$$

Next, a new input variable $v \in R^m$ is introduced as

$$v(k) = g(x(k), u(k), w) \quad (6.35)$$

or

$$y(k+r) = v(k) \quad (6.36)$$

where $g(x, u, w) = [g_1(x, u, w), ..., g_m(x, u, w)]^T$ and $y(k+r) = [y_1(k+r_1), ..., y_m(k+r_m)]^T$. Eqn. (6.36) obviously shows that the modified system is input-out linearized. Furthermore, if it is assumed that the m × m matrix

$$A(x, u, w) = \begin{bmatrix} \partial[h_1 \circ f^{r_1}(x, u)]/\partial u \\ \vdots \\ \partial[h_m \circ f^{r_m}(x, u)]/\partial u \end{bmatrix} \quad (6.37)$$

is nonsingular at (x^0, u^0), by the Implicit Function Theorem there exists a unique local solution of the nonlinear algebraic equation (6.35) as follows

$$u(k) = \alpha(x(k), v(k), w) \quad (6.38)$$

where $\partial \alpha / \partial v \neq 0$ at (x^0, v^0), $v^0 = g(x^0, u^0, w)$.

The nonlinear algebraic equation (6.35) may be solved using the numerical methods at each instant. In order to avoid the numerical complexity for solving the nonlinear algebraic equation (6.35), the right side term of Eqn. (6.35) is expended to the first-order term of the incremental term $\Delta u(k) = u(k) - u(k-1)$; that is,

$$v(k) = g(x(k), u(k-1), w) + A(x(k), u(k-1), w) \Delta u(k) + 0(\Delta^2 u) \quad (6.39)$$

Based on the nonsingularity of the matrix $A(x, u, w)$, an iterative algorithm is given as

$$u(k) = u(k-1) + \Delta u(k) \quad (6.40)$$

where
$$\Delta u(k) = [A(x(k), u(k-1), w)]^{-1} [v(k) - g(x(k), u(k-1), w)] \quad (6.41)$$

6.6 DYNAMIC LEARNING AND CONTROL OF UNKNOWN NONLINEAR PLANTS

6.6.1 Dynamic Learning Algorithm

Recently, several independent studies have found that the recurrent neural networks (RNNs) using the *dynamic backpropagation* (DBP) algorithm can approximate a wide range of input-output relationships of nonlinear systems to any desired degree of accuracy. For the supervised learning of the RNN, the purpose of the weight learning of the RNN is to estimate the weights such that the output of $y(k)$ of the RNN tracks the desired output $y_d(k)$ with an error which converges to zero as $k \to \infty$. Hence, if the weights of a RNN are taken into account as the unknown parameters of a nonlinear input-output system, the weight learning problem of the RNN can be phrased as a parameter identification procedure for the dynamic nonlinear system. As a matter of fact, a simple and natural extension of the BP algorithm for the multilayered feedforward network is the dynamic backpropagation (DBP) algorithm for the RNN. This learning approach was first studied by Williams and Zipser (1989), and Narendra and Parthasarathy (1991). A general DBP learning version will be derived in this subsection.

During such DBP learning sessions, the updating rule of the weights is given by

$$w(k+1) = w(k) + \eta \frac{\partial E(x)}{\partial w(x)} \quad (6.42)$$

where $w(k)$ is a estimation of the weight vector at time k, and η is a step size parameter, which affects the rate of convergence of the weights during learning. The output u_i of the network at current time k may be obtained only using the state and input of the network at past time $k - r_i$. The error index should be then defined as

$$E(k) = \tfrac{1}{2} \sum_{i=1}^{m} \left[y_{d,i}(k) - y_i(k) \right]^2 = \tfrac{1}{2} \sum_{i=1}^{m} e_i^2(k) \quad (6.43)$$

where $e_i(k) = y_{d,i}(k) - y_i(k)$ is a learning error between the desired and netwoprk output at time k. The partial derivatives of the error index $E(k)$ with respect to the weight of the network are obtained as follows using the dynamic neural model

$$\frac{\partial E(x)}{\partial w(x)} = - \sum_{i=1}^{m} e_i(k) \left[\frac{\partial y_i(k)}{\partial x(k)} \frac{\partial x(k)}{\partial x(k)} + \frac{\partial y_i(k)}{\partial w(k)} \right] \quad (6.44)$$

Let

$$z_{i,j}(k) = \frac{\partial x(k)}{\partial w_j(k)}, \quad j = 1, 2, ..., \bar{n} \tag{6.45}$$

where $\bar{n}$ is the total number of the weights which need to be adapted. Then, Eqn. (6.44) may be rewritten as

$$\frac{\partial E(k)}{\partial w_j(k)} = -\sum_{i=1}^{m} e_i(k) \left[\frac{\partial y_j(k)}{\partial x(k)} Z_j(k) + \frac{\partial y_j(k)}{\partial w(k)} \right] \tag{6.46}$$

where $Z_j(k) = [z_{i,1}(k), ..., z_{i,\bar{n}}(k)]$, and $z_{i,j}(k)$, $i = 1, 2, ..., \bar{n}$ are determined by the following time-varying linear system

$$z_{ij}(k) = \left[\frac{\partial f}{\partial x} \right] z_{ij}(k-1) + \left[\frac{\partial f}{\partial w_j} \right] \tag{6.47}$$

where

$$\left[\frac{\partial f}{\partial x} \right] = \frac{\partial f}{\partial x} [x(k-1), u(k-1), w(k)]$$

$$\left[\frac{\partial f}{\partial w_j} \right] = \frac{\partial f}{\partial w_j} [x(k-1), u(k-1), w(k)]$$

The partial derivatives in Eqns. (6.46) and (6.47) may be derived from Eqn. (6.27) where the first term in Eqn. (6.46) is due to the dynamic behavior of the RNN and the second term is a static partial derivative similar to that in conventional static back propagation algorithm (Rumelhart and McCelland, 1986).

6.6.2 Model Based Neural Control Schemes

In this subsection, the input-output linearized control technique combining the multilayered recurrent neural networks (MRNNs) is used to develop nonlinear adaptive control systems for unknown MIMO discrete-time nonlinear systems with on-line identification and control ability.

Consider a general class of unknown multi-input and multi-output (MIMO) discrete-time nonlinear system of the form

$$\begin{cases} x_p(k+1) = f_p(x_p(k), u(k)) \\ y_p(k+1) = h_p(x_p(k)) \end{cases} \quad (6.48)$$

where x is an n-dimensional state vector, u is an m-dimensional control vector, and $y \in R^m$ is a m-dimensional output vector. The mapping f_p and function h_p are assumed to be unknown. The problem of producing an output, irrespective of the initial state of the unknown nonlinear system, that converges asymptotically to a given reference output $y_m(k)$ will now be investigated. The reference output is not just a fixed function of time, but the output of a reference model, which is subject to some input $r(k)$ described by the equations of the form

$$\begin{cases} x_m(k+1) = Ax_m(k) + Br(k) \\ y_m(k) = Cx_m(k) \end{cases} \quad (6.49)$$

where $x_m \in R^m$, $r \in R^m$, and $y_m \in R^m$ are the state, input, and output of the model, respectively, A is a $n_m \times n_m$ Hurwitz matrix, and B and C^T are $n_m \times m$ vectors.

For the unknown nonlinear system (6.48), the design procedure of the learning control system is divided into the following two steps. The recurrent neural networks (RNN) with the weight vector w is first used to approximate the nonlinear plant (6.48). Then, Eqn. (6.48) may be governed by using the RNN as follows:

$$\begin{cases} x(k+1) = f(x(k), u(k), w(k)) \\ y(k) = h(x(k), w(k)) \end{cases} \quad (6.50)$$

where x is the state of the RNN, u is the input of the RNN, y is the output of the RNN, and $w(k)$ is a estimation of the weight vector of the RNN at instant k.

Next, assume that an nonlinear control law $u(k)$ is designed based on the equivalent control concept and the dynamic neural system (6.50) such that the output $y(k)$ of the dynamic neural system (6.50) will track asymptotically the output $y^m(k)$ of the reference model (Jin, Nikiforuk, and Gupta, 1993c); that is,

$$\lim_{k \to \infty} (y^m(k) - y(k)) = 0, \quad (6.51)$$

On the other hand, for learning the control scheme shown in Figure 6.12, the error used to train the RNN is defined as

$$e^*(k) = y_p(k) - y(k) \quad (6.52)$$

where $y_p(k)$ and $y(k)$ are the outputs of the neural network and the plant, respectively.

As shown in Figure 6.12, the error between the output of the model and the plant satisfies

$$e(k) = |y_p(k) - y^m(k)|$$

$$\leq |y_m(k) - y(k)| + |y_p(k) - y(k)| \qquad (6.53)$$

where the output $y(k)$ of the RNN tracks asymptotically the output $y^m(k)$ of the model by means of the learning control law $u(k)$. Hence, if the output of neural network RNN is trained to approximate the output of the plant with $\lim_{k\to\infty} e^*(k) = 0$, the output of the plant is then adaptive controlled to track asymptotically the output $y^m(k)$ of the model; that is $\lim_{k\to\infty} e(k) = 0$. In fact, the recurrent network RNN is used to identified the nonlinear plant on-line, while the control law is constructed based on the identification results of the neural network.

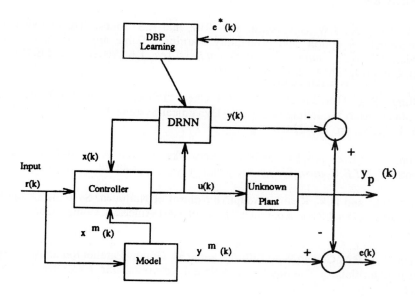

Figure 6.12. Dynamic learning control of a unknown plant using RNN

6.7 CONCLUSION

Neural structures of both the static multilayered feedforward neural networks (MFNNs) and the dynamic recurrent neural network, the two main classes of artificial neural networks, have been discussed in this chapter with respect to the learning and adaptive control of unknown, complex nonlinear systems. A comparison of the neural network based schemes discussed in the chapter and conventional self-turning and model reference adaptive systems shows that the former needs less *a prior* knowledge about the unknown plant, and that the structure and the mathematical expressions of the former are also simpler. Indeed, since the nonlinear plant is assumed to be unknown, and modeled on-line by the neural network, the neural network based control systems have the good robustness for the time-varying properties of the plant. These advantages are difficult to achieve using the conventional adaptive control systems. Some simulation studies can be found in Narendra and Parthasarathy (1990, 1991), and Jin, Nikiforuk and Gupta (1992, 1993c).

Furthermore, except for the problem of the theoretical analysis of the convergence of the weight learning algorithm, and the stability of the adaptive control schemes, an interesting and important aspect of the learning and control algorithm in the future will be to develop a generalized nonlinear learning controller or a universal learning controller which can be applied to an arbitrary unknown nonlinear dynamic system through an on-line learning process.

REFERENCES

1. Anderson, J.A., J.W. Silverstein, S.A. Ritz, and R.S. Jones. "Distinctive features, categorical perception, and probability learning: Some applications of a neural model." *Neurocomputing: Foundation of Research*. J.A. Anderson and E. Rosenfeld, Eds. Cambridge, MA: MIT Press (1988).

2. Billings, S.A., H.B. Jamaluddin, and S. Chen. "Properties of Neural Networks with Applications to Modelling Non-linear Dynamical Systems." *Int. J. Control*. Vol. 55, No. 1 (1992), pp. 193-224.

3. Chen, F.C. "Back-propagation Neural Networks for Nonlinear Self-Tuning Adaptive Control." *IEEE Control System Magazine*. (1990), pp. 44-48.

4. Chen, F.C. and H.K. Khalil. "Adaptive Control of Nonlinear Systems Using Neural Networks-A Dead-Zone Approach." Proc. of 1991 American Control Conference, pp. 667-672.

5. Chen, S., C.F.N. Cowan, S.A. Billings, and P.M. Grant. "Parallel Recursive Prediction Error Algorithm for Training Layered Neural Networks." *Int. J. Control*, Vol. 51, No. 6 (1990), pp. 1215-1228.

6. Grossberg, S. "Nonlinear neural networks: Principles, mechanisms and architectures." *Neural Networks*. Vol. 1, No. 1 (1988), pp. 17-61.

7. Hecht-Nielsen, R. "Theory of the Back-propagation Neural Network." Proc. Int. Joint Conf. on Neural Networks (1989), pp. I-593-605.

8. Hernandez, E. and Arkun, Y. "Study of the Control-Relevant Properties of

Backpropagation Neural Models of Nonlinear Dynamical Systems." *Computer Chem. Engng.*, Vol. 6, No. 4 (1992), pp. 227-240.

9. Hunt, K.J. and D. Sbarbaro. "Neural Networks for Nonlinear Internal Model Control." *IEE* Proc. D, Vol. 138, No. 5 (1991), pp. 431-438.

10. Hunt, K.J., D. Sbarbaro, R. Zbikowski, and P.J. Gawthrop. "Neural Networks for Control Systems - A Survey." *Automatica.* Vol. 28, No. 6 (1992), pp. 1083-1112.

11. Hui, S. and S.H. Zak. "Dynamical Analysis of the Brain-State-in-a-Box (BSB) Neural Models." *IEE Trans. on Neural Networks.* Vol. 3, No. 1 (1992), pp. 86-94.

12. Hush, D.R. and B.G. Horne. "Progress in Supervised Neural Networks." *IEEE Signal Processing Magazine.* No. 1 (1993), pp. 8-39.

13. Isermann, R. *Digital Control Systems.* 2nd ed., Vol. 1 (1989), Springer-Verlag.

14. Isidori, A. *Nonlinear Control System.* Springer Verlag, New York (1989).

15. Jin, L., P.N. Nikiforuk, and M.M. Gupta. "Adaptive Tracking of SISO Nonlinear Systems Using Multilayered Neural Networks." Proc. of the 1992 American Control Conference, Vol. 1 (1992), pp. 56-60.

16. Jin, L., P.N. Nikiforuk, and M.M. Gupta. "Direct Adaptive Output Tracking Control Using Multilayered Neural Networks." *IEE* Proc. D. (1993a).

17. Jin, L., P.N. Nikiforuk, and M.M. Gupta. (1993b). "Adaptive Control of Discrete-Time Nonlinear System Using Multilayered Recurrent Neural Networks." *IEE* Proc. D [1993b, in press].

18. Jin, L., P.N. Nikiforuk, and M.M. Gupta. "Multilayered Recurrent Neural Networks for Learning and Control of Unknown Nonlinear Systems." *J. of Dynamics, Measurement Control.* [1993c, in press].

19. Michel, A.H., J. Si, and G. Yen. "Analysis and Synthesis of a Class of Discrete-Time Neural Networks Described on Hypercubes." *IEE Trans. on Neural Networks.* Vol. 2, No. 1 (1991), pp. 32-46.

20. Miller, III, W.T., R.S. Sutton, and P.J. Werbos, Ed. "Neural Networks for Control." The MIT Press, London (1990).

21. Mukhopadhyay, S. and K.S. Narendra. "Disturbance Region in Nonlinear Systems Using Neural Networks." *IEEE Trans. on Neural Networks.* Vol. 4., No. 1 (1993), pp. 63-72.

22. Narendra, K.S. and K. Parthasarathy. "Identification and Control of Dynamical Systems Using Neural Networks." *IEEE Trans. Neural Networks*, Vol. NN-1, No. 1 (1990), pp. 4-27.

23. Narendra, K.S. and K. Parthasarathy. "Gradient Methods for the Optimization of Dynamical Systems Containing Neural Networks." *IEEE Trans. Neural Networks*, Vol. 2, No. 2 (1991), pp. 4-27.

24. Nijmeijer, H. and A. J. Van der Schaft. *Nonlinear Dynamical Control Systems*. Springer-Verlag, New York Inc. (1990).

25. Parlos, A., A. Atiya and K. Chong. "Recurrent Multilayer Perceptron for Nonlinear System Identification." *Proc. IJCNN*. Vol. II (1991), pp. 537-540.

26. Psaltis, D., A. Sideris, and A.A. Yamamura. "A Multilayered Neural Network Controller." *IEEE Control System Magazine (1988)*, pp. 17-21.

27. Rumelhart, D.E. and J.L. McCelland. "Learning Internal Representation by Error Propagation." *Parallel Distributed Processing: Exploration in the Microstructure of Cognition*. Vol. 1: Foundations, MIT Press (1986).

28. Sanner, R.M. and J.J.E. Slotine, J.J.E. "Gaussian Networks for Direct Adaptive Control." *IEEE Trans. Neural Networks*. Vol. 3, No. 6 (1992), pp. 837-863.

29. Sanner, R.M. and J.J.E. Slotine. "Stable Adaptive Control and Recursive Identification Using Radial Gaussian Network." *Proc. of the 30th CDC*. Vol. 2 (1991), pp. 2116-2123.

30. Simpson, P.K. *Artificial Neural Systems*. Pergamon Press (1990).

31. Sira-Ramirez, H. (1991). "Nonlinear Discrete Variable Structure Systems in Quasi-Sliding Mode." *Int. J. Control*. Vol. 54, No. 5 (1991), pp. 1171-1187.

32. Sudharsanan, S.I. and M.K. Sundereshan. "Training of a Three-Layer Dynamical Recurrent Neural Network for Nonlinear Input-Output Mapping." *Proc. IJCNN*. Vol. II (1990), pp. 111-115.

33. Tzirkel-Hancock, E. and R. Fallside. "Stable Control of Nonlinear Systems Using Neural Networks." *Int. J. of Robust and Nonlinear Control*. Vol. 2, No. 1 (1992), pp. 63-86.

34. Widrow, B. and L.A. Lehr. "30 Years of Adaptive Neural Networks: Perceptron, Madaline, and Backpropagation." *Proc. IEEE*. Vol. 78, No. 9 (1990), pp. 1415-1441.

35. Williams, R. and D. Zipser. "A Learning Algorithm for Continually Running Fully Recurrent Neural Networks." *Neural Computation*. Vol. 1 (1989), pp. 270-280.

7

THE APPLICATION OF ARTIFICIAL NEURAL NETWORKS IN EDITING NOISY SEISMIC DATA

X. Zhang and Y. Li, Tsinghua University
Beijing, CHINA

People have gained great profits from the application of computer technologies in geophysical exploration, but there are still labor intensive areas where human intelligence is needed. Noisy seismic data editing is one of these. It is a special two-class pattern recognition problem. An automatic editing approach is discussed in this chapter. It applies a neural network model called a novelty filter. The learning procedure uses only good trace samples, and can even be implemented in an unsupervised way. Experimental results show that the approach is a promising one.

7.1 INTRODUCTION

Computer technologies nowadays are widely applied in the field of geophysical exploration, but there are still many tasks of seismic signal processing and discriminating that must be fulfilled by human processors or interpreters. Those tasks require human expertise and know-how, and the results are somehow subjective. This is work where not only computing but intelligence is also needed. There have been numerous attempts to develop methods of performing some of these tasks by computers. The recent developments in artificial intelligence and artificial neural network technologies suggest some new ideas. In this chapter, a new method of automatically editing noisy seismic data is discussed. It is fulfilled by an artificial neural network model called a novelty filter.

Noise is a serious problem in seismic signals as well as in other signals. A lot of seismic processing procedures, such as stacking and filtering, are aimed at reducing noises. However, in field data there are some traces that are too noisy, that is, they are dominated or partially dominated by noise, so that no noise-reducing processings perform well on them. These traces or the noisy parts of these traces, called bad traces, must be removed or muted (replaced by zeros) from the data set at the beginning of processing, for their existence will adversely affect the quality of the whole processing. This procedure is usually called editing.

In most production lines at present, editing is usually done by human processors using paper plots or workstations. Because of the vast size of modern seismic data, especially 3-D seismic data, manual editing is very labor consuming. This is one of the few remaining labor intensive areas in seismic data processing, and the quality of manual editing depends heavily on the skill and working condition of the human processor.

Although the goal of trace editing is to maximize the S/N ratio (SNR) of the stacked data in the zone of interest, the editing task is more an interpretive process than a computing one, for no quantified way of estimating the SNR of the data is available. As for a human editor, he performs simply a recognition procedure. Decisions are made according to his knowledge, rules of thumb, and visual comparison.

It is now clear that editing is in fact a pattern recognition task, and is a two-class problem, with one class of good traces and one of bad traces. But it is not as easy as it looks, for it is hard to extract features to describe the two classes. Bad traces themselves are usually of many different types, the differences of which are no less than the difference between good and bad traces. There are people who have summarized several main types of noises and designed methods to recognize them one by one.[2] In this way, a lot of computer time is needed, and the quality of the results depends on whether the summary is good and general.

Artificial neural networks have shown great potential in many pattern recognition problems, especially in problems where rules cannot be found and features are hard to be extracted. We have developed an approach using a simple neural network model, the novelty filter model, to solve the difficulties in trace editing. After proper "learning," our novelty filter editor can easily distinguish bad traces of every types from good ones. The learning procedure uses only some good trace samples, and even can be implemented unsupervised in most real cases in which good traces overwhelm bad ones by numbers. Experiments on both synthetic data and field data show that our approach is a robust one and is quite efficient.

7.2 ARTIFICIAL NEURAL NETWORKS AND NOVELTY FILTER

Artificial neural network models or simply "neural nets" have been studied for many years with an aim to achieve human-like performance in the field of pattern recognition and other artificial intelligence fields. There are already many models proposed (e.g. the models overviewed by Lippmann[3]). All these models attempt to achieve their goals via dense

The Application of Artificial Neural Networks in Editing Noisy Seismic Data 155

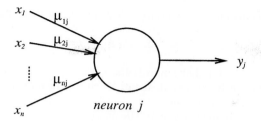

Figure 7.1. A Basic Neuron.

interconnection of simple computational elements, or "neurons," and the connection weights are typically adapted during use or training. Some neural nets have already shown remarkable performances in areas such as speech and image recognition.

A basic network unit or a neuron is illustrated in Figure 7.1. Let $X = [x_1, x_2, \cdots, x_n]^T$ denote the input vector and $W = [\mu_{1j}, \mu_{2j}, \ldots, \mu_{nj}]^T$ denote the weight vector, then the output y_j can be written as

$$y_j = \sigma_j \left[W_j^T X - \theta \right] = \sigma_j \left[\sum_{i=1}^{n} \mu_{ij} x_{i-\theta_j} \right],$$

where $\sigma_j(\cdot)$ is some kind of transfer function and θ_j is a threshold or offset.

In adapting the weights, there are many learning rules, most of which bear the form

$$dW_j \big/ dt = \phi(y_i) X - \gamma(y_i) W_j$$

where $\phi(\cdot)$ and $\gamma(\cdot)$ are certain scalar functions.

As a special case, we can build our network as illustrated in Figure 7.2, with n neurons connected with each other by feedbacks. Let the transfer function $\sigma_j(\cdot)$ be linear and let $\phi(y_j) = -\alpha y_j$, ($\alpha > 0$) while $\gamma(y_j) = 0$, we get the special neural network model called novelty filter. The output function is

$$y_j = x_j + \sum_{i=1}^{n} \mu_{ij} y_i, \qquad j = 1, 2, \ldots, n$$

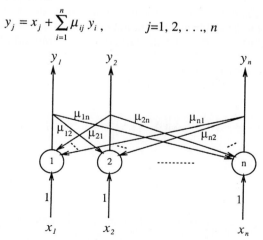

Figure 7.2. Novelty Filter Structure.

and the feedback weights are adapted during learning under the following rule,

$$d\mu_{ij}/dt = -\alpha\, y_i\, u_j, \quad i=1, 2, \ldots, n \quad j=1, 2, \ldots, n$$

where $\alpha > 0$ is a gain factor that controls the speed and convergence of the learning procedure. This learning rule can be called anti-Hebb-rule or minus-Hebb-rule for it is just the negative of the famous Hebb-rule.

The novelty filter is a dynamic system. The mathematical aspect of its performance was fully discussed in[1] by Kohonen. Intuitively we can see that if any input pattern X applied to a novelty filter is held stationary, then through the adaptation of the weights, the output will monotonically tend to zero. What is really interesting and useful is that such an asymptotic balance can simulaneously be achieved for many different input patterns, the number of which* is less than n, the number of input terminals. The training order of the patterns can be arbitrary, as long as each pattern appears as input for adequate time. After training, the network will give only a small output when the input vector is similar with one of the trained patterns or some linear combination of the trained patterns. On the other hand, if the input pattern is a novel one, then a large output will be obtained from the filter, the amplitude of which represents the degree of dissimilarity with the trained patterns. This is just like the psychologic phenomenon of habituation. The novelty in the input can be detected out according to the output value. That is why the network is called a novelty filter.

The above property can be mathematically proven and the convergence of the system is guaranteed if the α and the initial μ_{ij} are properly chosen.[1] Let M be the weight matrix that $M=[\mu_{ij}]$, X and Y be the input and output vector, then the system equation can be written as

$$Y = X + M\,Y,$$

which derives

$$Y = (I-M)^{-1} X = \Phi\, X,$$

where I is the unit matrix and $\Phi = (I-M)^{-1}$. We use M_0 and Φ_0 to denote the initial value of M and Φ. For the system to be stable and the training procedure to converge, M_0 should be chosen so that $(I-M_0)^{-1}$ exists and Φ_0 is a symmetric positive semidefinite matrix. The gain factor α should be

$$\varepsilon < \alpha < \lambda - \varepsilon,$$

with

and

$$\lambda = \left(X^T \Phi_0^3 X\right)\left(X^T \Phi_0^2 X\right) - 1\left(X^T \Phi_0^4 X\right)^{-1},$$

$$0 < \varepsilon < \lambda.$$

---

*Strictly speaking, it should be the number of orthogonal patterns contained in the input pattern set. But the input patterns need not be orthogonal to each other. In our experiments, many training samples we chose are usually correlated with each other. In this way, the resulted novelty filter can be less sensitive to small amplitude or phase noise.

Kohonen[1] showed that under the above conditions, the procedure converges monotonically. This turns out to be a little complicated. In practical cases, we can just choose M_0 to be zero matrix or some small-random-valued symmetric matrix. If the input vectors are normalized by their modules before being feed into the network, the α can be chosen as $0 < \alpha < 1$, with the specific value adjusted by trial-and-error to achieve better performances.

The novelty filter can be applied in many problems. It can be seen as a special two-class pattern classifier that uses samples from only one class for training. This property is especially useful in cases where one cannot get enough typical samples of both classes. Kohonen has used this novelty filter to find out novel points in his input images. As for its application to seismic trace editing, our idea is to let the network habituate with good traces and then decide whether a trace is bad or good according to whether it is detected as novel or not.

7.3 NOVELTY FILTER FOR TRACE EDITING: SUPERVISED

As we have mentioned in the introduction, human processors perform an interpretive procedure when editing. One can easily see that, although noisy seismic traces are very different from each other, they are all different from good traces, which is the only "feature" they have in common. Or simply put, bad traces all don't look like good ones. On the other hand, good traces are relatively more similar to each other, at least in the same gather. Based on this observation, we designed our approach by first making the novelty filter habituated with good traces and then recognizing the novel input as a bad trace.

In many practical cases, only some parts of a noisy trace but not the entire trace needs to be removed. So we divide a seismic trace into several time gates, the length of which can be easily adjusted. The typical time gate we used in experiments is of 100 sample points long, or 200 ms long if the sample rate is 2 ms. This can be varied depending on the situation, but if the gates are too long, editing cannot be very accurate, and if too short, there may be more chances for mis-recognition. (In this chapter, unless specifically indicated, the word "trace" refers to a certain time gate of a trace.) We use the seismic amplitude series in a time gate as the input vector X of our novelty filter. Figure 7.3 is the structure of the novelty filler editor, where a module block is added to the output vector Y to get a scalar output. In our experimental programs, the input vectors are normalized by their averaged absolute amplitudes, which makes that the α should be much smaller than when the vectors are module-normalized.

The working of the novelty filter editor is divided in two phases, one learning or training phase and one recognizing phase. Figure 7.4 diagrams the procedure. In the learning phase, we choose some samples from good traces to train the network so that it gets habituated with good traces by the end of this phase. In the recognizing phase, the network "scans" over the whole data set trace by trace and time gate by time gate. Whenever the data in certain time gates are dissimilar with good traces, the network will give a large output. We decide a threshold, so that the trace whose corresponding output exceeds the threshold is recognized as a bad trace and is then removed. Since bad traces usually are quite different from good ones, and since the training procedure is monotonous when α is proper, we practically never needs to wait for the procedure to converge. In fact, our experiments show that just a few rounds of training is enough for most bad traces to be distinguished.

Figure 7.5 is one of our experimental examples. In this example, the noises are synthetic while the normal data are of a real field CDP gather. We used the data in the two boxes shown

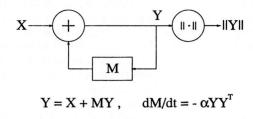

$$Y = X + MY, \quad dM/dt = -\alpha YY^T$$

Figure 7.3. Novelty Filter Editor Structure.

in the left section of Figure 7.5 as learning samples. The right section is the editing result, from which we can see that all the noisy parts are removed while no normal data are mis-removed. The network was trained 10 rounds with each of the 20 good trace samples with α-0.002.

Figure 7.6 is the output histogram of this example, with the horizontal axis being the output value and the vertical axis being the corresponding appearing frequency. From this histogram, we can see that the good and bad traces are well separated and the classification threshold can be chosen easily with the help of this histogram.

7.4 NOVELTY FILTER FOR TRACE EDITING: UNSUPERVISED

The above method used supervised learning, that is, human beings provided the network with some learning samples, and the network made its decision according to what it had "learned" from the samples. In fact, the learning can also be done unsupervisedly, and then

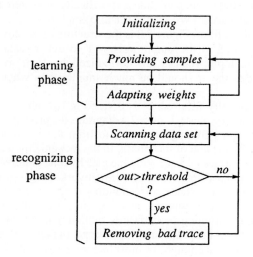

Figure 7.4. Editing procedure under Supervised Learning.

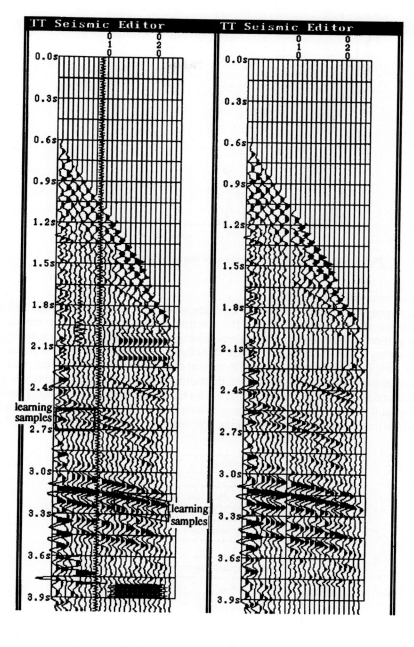

Figure 7.5. An editing example with Supervised Learning.

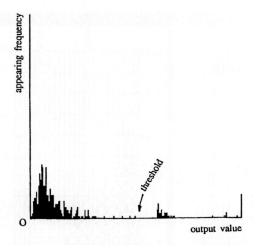

Figure 7.6. The Output Histogram of the example in Figure 5.

the editing becomes fully automatic.

We may agree that, in most cases if not all cases, noisy traces that must be removed are only the minorities in real seismic data sets. If this is true, then we can use all the traces (from one or a few gathers) as learning samples without selection. The procedure is then as in Figure 7.7 In the learning phase, the network "scans" over the whole data set, thus, we can use all the traces as samples to train the network. Since most of the samples are good ones, and the few bad ones are likely to be of different types, then after the learning phase, the network will become habituated with the good ones, but not or less habituated with the bad ones. Thus the bad traces can still be distinguished in the recognition phase. In Figure 7.8 are two examples that were edited using this unsupervised leaning procedure. These are all field data, and their

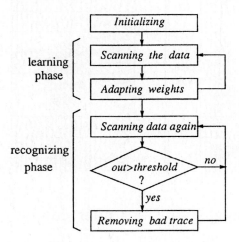

Figure 7.7. Editing procedure under unSupervised Learning.

SNRs are low, but the editing results are still quite acceptable. No other preprocessing procedure has been done to these data sets except the amplitude recovery. The time gate used in these examples is also 100 points or 200ms long. We used our Sun4/280 workstation (10 MIPS, 32MB memory) to compute these examples, and one of these

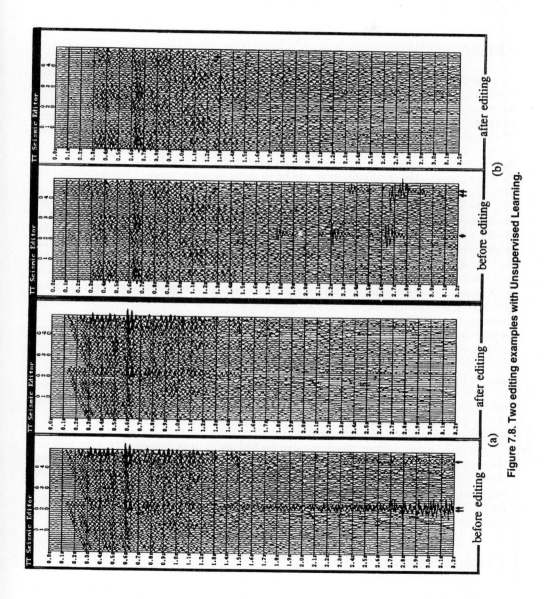

Figure 7.8. Two editing examples with Unsupervised Learning.

two experiments cost about 180s CPU time including the time used for data reading and writing.

In our experiment, most (more than 90%) of the computing time is spent on the learning phase. On production lines, if the gathers in a line are similar to each other, then we can use just a few gathers for the learning phase, leaving the rest of the gathers only for the recognition phase. In this way, a lot of computer time can be saved.

7.5 DISCUSSION

The conventional artificial intelligence approach for solving intelligence-involved tasks is to summarize some features and rules according to experts' analysis, experiences, and knowledge, then write the rules in programs for the computer to make decisions. As a typical example, Anderson & McMechan[2] applied this idea for the seismic trace editing problem. They summarized several main types of noisy traces that must be removed and developed rules to detect them one by one. The performance of this kind of method depends heavily on how well the types and rules are summarized and how well they suit the specific case at hand. In some complicated situations, there can be so many different types of bad traces that the idea of detecting them with rules one by one turns out to be impractical.

The idea is quite different when artificial neural networks are applied. Data samples are used to train the network and "decision rules" are not written into the program but "learned" by the networks themselves from the training data. Performance no longer depends on the expertise the designer used, and the same network can work well in quite different cases, for one needs only to use different training samples for different situations. McCormack[5] has proposed a neural network approach which used the back propagation (BP) network. He used FFT amplitudes of the seismic trace and 5 other seismic features as the pattern vectors. It was a supervised method in which both bad trace and good trace samples must be provided for training, and the editing was based on entire traces. In comparison, our novelty filter editor is based on time gates so it can do accurate editing. It is a great advantage that the training of novelty filters needs only good trace samples, especially in cases where there are too many kinds of bad traces for one to give typical samples of each kind. In most practical cases, the novelty filter can even be implemented in an unsupervised way, thus reducing human effort.

One limitation of our method is that the correlations between neighboring traces have not been considered, thus some bad traces with a wrong phase cannot be distinguished. This is also the limitation of many other automatic editing methods. Research work aimed at solving this problem is already in progress.

It should be pointed out that to remove the bad traces from data sets is not the best way to deal with strong noise, for there are maybe still some useful information in the data being removed, although perhaps very weak. The ideal way would be by filtering the noise out while preserving the useful information, or using some "proper" data but not zeros to replace the bad traces. Researchers presently have not found such a way as yet and the editing method is still the cheapest and most practical way available.

7.6 CONCLUSION

In this paper, we have discussed an artificial neural network approach for automatically editing noisy seismic data. The network model used is the simple novelty filter model, a single-layered feedback neural network with feedback weights adapted during learning according to the anti-Hebb-rule. The network can get habituated with the training patterns and does not respond or respond only slightly to input patterns that are similar to the trained patterns. However, large response will be obtained if the input is a novel one, thus the input patterns can be divided into two classes. This property is applied for the seismic trace editing problem, which is a special two-class pattern recognition problem. Seismic amplitude series in time gates are used as input vectors, and only good trace samples are used for training. For most practical cases, the novelty filter editor can even work in an unsupervised way, which reduces more human effort. Experiments on both synthetic and field data show that the approach works well.

REFERENCES

1. Kohonen, T. *Self-Organization and Associative Memory* (2nd edition). Springer-Verlag, Berlin (1988).

2. Anderson, R.G. and G. A. McMechan. "Automatic Editing of Noisy Seismic Data." *Geophysical Prospecting.* V.37, No.8 (1989), pp.875-892.

3. Lippmann, R.P. "An Introduction to Computing With Neural Nets." *IEEE ASSP Magazine (*April, 1987), pp. 4-22.

4. Zhang, X. and Y. Li. "Automatic Editing of Noisy Seismic Data Using an Artificial Neural Network Approach." *Acta Geophysica Sinica* (in Chinese). V.35, No.5 (1992), pp. 637-643.

5. McCormack, M.D. "Seismic Trace Editing and First Break Picking Using Neural Networks." *Extended Abstracts of SEG 60th Annual Meeting*(1990), pp. 321-324.

8

FOUNDATIONS OF FUZZY NEURAL COMPUTATIONS

M. M. Gupta and H. Ding
University of Saskatchewan
Saskatoon, Saskatchewan, CANADA

In this chapter, some basic notions and fundamental principles of fuzzy-neural computations are presented, thus laying a foundation for fuzzy neural networks. A geometric interpretation of fuzzy-neuronal operations is first presented, and then followed by an intuitive discussion of neuro-fuzzy computational logic. A generalized morphology of computational neurons is also presented which uncovers some of the common characteristics of the information processing aspects of the currently-used neuronal models. Two forms of neuronal morphologies, namely *Product-Summation-Monotonicity* (*PSM*) and *Difference-Summation-Radial* (*DSR*) are discussed. These concepts are then extended to fuzzy neurons, and their intrinsic properties are examined.

8.1 INTRODUCTION

From the viewpoint of neuronal computations, current studies on neural networks can be classified into three main kinds of research subjects. The first one is on the modeling of a single neuron, which can be modeled as a static neuron, a dynamic neuron or a domain-specific neuron. The second one is on the architecture of a network of neurons with layered feedforward networks (static) or recurrent (feedback) dynamic networks. The third one is on the operational aspects of neuronal computations, in which fuzzy (operational) logic and fuzzy arithmetic have been integrated with neural computations to enhance the uncertain information processing capabilities of neural networks. This integration yields a new class of neural networks, known as *fuzzy neural networks*(FNNs).[1,2] Many kinds of architectures of FNNs have been proposed in the current literature. Some of them have successfully been applied to some real-life problems. However, a fundamental problem in current studies on fuzzy-neural computations is the lack of biological-like computationality and flexibility of neuro-fuzzy logic for the design of fuzzy neural models. The overwhelming volume of research and publications often puzzle the novices in this interdisciplinary field of fuzzy logic and neural networks.

In this paragraph, a brief survey of some of the past studies in this field is presented. Kuncicky and Kandel[3] proposed a fuzzy neuron in which the output of one neuron is represented by a fuzzy level of confidence. Ishibuchi[4] and Cohen[5] used a learning algorithm for the interval vector to develop an architecture of fuzzy neural networks. They mapped an n-dimensional fuzzy input vector into a scalar fuzzy number using the neural networks where the activation function was extended to a fuzzy input-output relation by the extension principle. Yamakama[6] proposed a fuzzy neuronal model and applied it successfully to a pattern recognition problem, in which the synaptic weights were designed using the fuzzy membership functions, and the aggregation operation was performed using the fuzzy MIN-operation.

Generally speaking, current studies on fuzzy-neural computations can be classified into two kinds of approaches. In the first fuzzy-neural approach, the functionality of the system of fuzzy logic can be approximated by using neural models in which ordinary arithmetic is employed in neural computations. Most learning algorithms used in conventional non-fuzzy neural networks are appropriate for the design of such models. Using this approach, membership functions of fuzzy sets and fuzzy if-then rules can be acquired and adapted. This approach has been widely applied in areas such as fault diagnosis,[7] pattern recognition[8] and process control.[9]

In the second fuzzy-neural approach, fuzzy logic is employed in the design of fuzzy neuronal morphologies. In this approach, several *ad hoc* forms of fuzzy neurons have been proposed to approximate problems with fuzzy uncertainties in a way that they are structurally similar to what is found in a fuzzy logic system, such as fuzzy neuronal models for fuzzification, fuzzy reasoning and defuzzification. However, little progress has been made in this direction so far. More importantly, only a few original studies on the mathematical morphology and the operational mechanism of a fuzzy neuron have so far been reported.[10,11]

In this chapter, an intrinsic interpretationis presented of neuronal computational mechanisms, followed by some basic concepts of fuzzy-neural computations based on these mechanisms. First, the mechanisms of fuzzy neuronal operations are presented geometrically using a fuzzy-set-as-a-vector viewpoint. This can be considered as being fundamental to neuro-fuzzy logic which is a type of computational *soft-logic* for FNNs. The notion of fuzzy space of fuzzy vectors and its algebra are presented in the Appendix. Based on this geometric presentation, and from the viewpoint of the mechanism of neuronal information processing, a generalized morphology of a computational neuron is presented and discussed in Section 8.2. This is followed by a discussion of a generalized morphology of fuzzy neurons in Section 8.3. Three kinds of computational fuzzy neural networks are present and their intrinsic properties examined in Section 8.4. Finally, conclusions and suggestions for the future work in this area are discussed in Section 8.5.

8.2 NEURONAL INFORMATION PROCESSOR: SOME BASICS

In this section the generalized morphology of a computational neuron is first presented from the information processing point of view. Two fuzzy neuronal models are then described and compared from both the computational and the biological viewpoints.

8.2.1 A Generalized Morphology of Computational Neurons

From the neuro-biological, as well as from the neuro-mathematical viewpoint, two key operational elements can be identified in a neuron:[10] the *synapse* and the *soma*. These elements are responsible for providing two kinds of mathematical operations in a neuro-information processing mechanism: the *synaptic operation* and the *somatic operation*. A generalized mathematical morphology of a computational neuron is shown in Figure 8.1a.

First, consider the mathematical operations of a single neuron, which may be fuzzy or non-fuzzy. A neuron, as an *information processor*, receives a neuronal input (stimuli) vector $x(t) = (x_1(t), x_2(t), \ldots, x_n(t))^T R^n$, conducts the synaptic operation between the neuronal input vector $x(t)$ and the synaptic weighting vector $w(t) = (w_1(t), w_2(t), \ldots, w_n(t))^T \in R^n$, sends the weighted synaptic signal $u(t) \in R^n$ to the soma via dendrites for the somatic processing (aggregation, thresholding and nonlinear activation) that follows, and finally yields a scalar axonic output $y(t) \in R$. In the learning mode, the synapse (including the threshold in the neuron) learns by continuously adapting its strength to the new neuronal inputs $x(t)$ by an appropriate learning algorithm. These neuronal operations are described as follows, Figure 8.1a:

1) Synaptic operation:

$$u_i(t) = x_i(t) \; @ \; w_i(t), \quad i=1,2,\ldots,n \tag{8.1}$$

2) Somatic operations:

(i) Aggregation:

$$v(t) = \oiint u_i(t), \quad v(t) \in R \tag{8.2}$$

(ii) Thresholding:

$$z(t) = v(t) \; \Xi, \quad z(t) \in R \tag{8.3}$$

(iii) Nonlinear activation:

$$y(t) = f[z(t)], \quad y(t) \in R \tag{8.4}$$

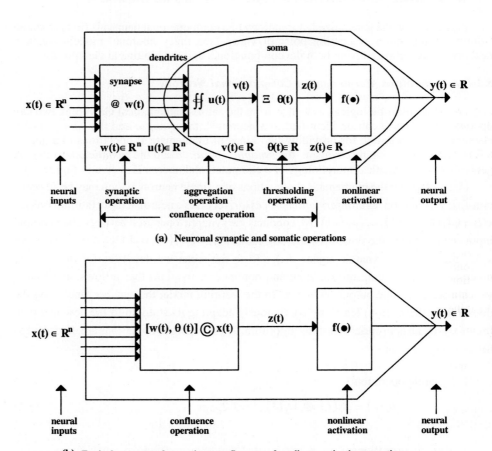

Figure 8.1 A generalized mathematical morphology of a simple computational neuron: $x(t) \in R^n$: neuronal inputs, $w(t) \in R^n$: synaptic weights, $\theta(t) \in R^n$: threshold, $z(t) \in R$: measure of mutual relationship, $y(t) \in R$: neuronal output, @: synaptic operator, ⨍: aggregation operator, Ξ: thresholding operator, ©: confluence operator, $f[\cdot]$: nonlinear activation function.

where '@', '⨏', and 'Ξ' are the generalized neuro-computational operators representing respectively the synaptic, aggregation and thresholding operations, which reflect a kind of operational logic for neuronal computations, $\theta(t) \in R$ is the threshold (bias), and $f[\cdot]$ is a neuronal nonlinear activation function.

The main objective of the neuronal operations is to provide a nonlinear and adaptive (with learning) mapping operation of the neuronal input stimuli vector $x(t) \in R^n$ into a scalar output $y(t) \in R$. The adaptive ability is ensured by the process of self-adjusting the synaptic weights $w(t)$ by using an appropriate learning algorithm. The somatic nonlinear activation $f[\cdot]$ enhances the flexibility and robustness of the neuronal computation under uncertainties. The synaptic weighting vector $w(t)$ can be viewed as a representation of *accumulated learned experiences (knowledge)* stored in the synapses of a neuron.

The main neuronal computational mechanism consisting of the synaptic and the aggregation operations, is to measure and detect the *mutual relationship** between the fresh neuronal input vector $x(t)$ and the strength of the synaptic weighting vector $w(t)$ (which represents the accumulated past experiences). That is, if there is a mutual relationship (similarity or dissimilarity) between $x(t)$ and $w(t)$, it will be detected and measured by the synaptic and aggregation operations, yielding a scalar signal $v(t)$. If $v(t)$ exceeds a threshold $\theta(t)$, in some sense, a degree of the mutual relationship $z(t)$ will be extracted as a neuronal activation, and a graded output $y(t)$ will be yielded through a nonlinear activation function $f[\cdot]$.

The function of the thresholding operation corresponds to a bias to the activation value $z(t)$. The bias itself is related to the past knowledge $w(t)$. In order to handle complex practical problems, each neuron in a neural network usually needs to adjust the bias in its neuronal activation with respect to different requirements of physical and computational functions. Therefore, the threshold $\theta(t)$ should also be adapted along with the synaptic weighting vector $w(t)$. Mathematically, the following augmented vectors are defined to combine the thresholding operation with the synaptic and aggregation operations:

$$x_a(t) = (x_0(t), x_1(t), \ldots, x_n(t))^T \in R^{n+1}, \ x_0(t) = 1 \qquad (8.5a)$$

and

$$w_a(t) = (w_0(t), w_1(t), \ldots, w_n(t))^T \in R^{n+1}, \ w_0(t) = \theta(t) \qquad (8.5b)$$

where $x_0(t) = 1$ and $w_0(t) = \theta(t)$ correspond to the threshold (bias) term. Finally, a nonlinear activation function $f[\cdot]$ provides a scalar graded neuronal output $y(t) \in R$. The overall neuronal operations of a single neuron for $x(t) \in R^n$ to $y(t) \in R$, may, thus, be considered as a nonlinear mapping operation from many inputs to a single output, as shown in Figure 8.2.

From the viewpoint of information processing, the neuronal mapping operation of a single neuron can also be classified into two operational phases, Figure 8.1b:

i) *confluence operation,* and
ii) *nonlinear activation operation.*

---

*In principle, the mutual relationship between two objects x and y can be measured by an index of similarity or dissimilarity. The kind of measure used depends upon the specific applications.

Let the confluence operation be defined as an operation for the measure of the mutual relationship between the fresh neuronal input vector $x(t)$ and the synaptic weighting vector $w(t)$. Thus, the confluence operation is an integrated operation, consisting of the synaptic, aggregation and thresholding operations, defined as

$$z(t) = [w(t), \theta(t)] \copyright x(t) \qquad (8.6a)$$

where $\copyright$ is the confluence operator, and $z(t)$ may be called a measure of the mutual relationship. If the thresholding operation can mathematically be integrated into the synaptic and aggregation operations (see the discussion in Section 8.2) using augmented vectors defined in (5a-5b), the confluence operation can also be defined as:

$$z(t) = w_a(t) \copyright x_a(t) \qquad (8.6b)$$

Finally the nonlinear activation function $f[\cdot]$ grades the measured degree $z(t)$ to enhance the robustness and flexibility of the neuronal mapping.

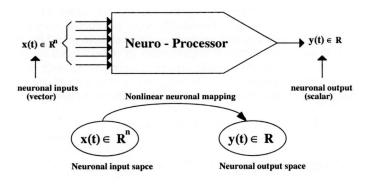

Figure 8.2. A single neuronal computation as a nonlinear mapping from $x(t) \in R^n$ to $y(t) \in R$.

The problem now is how to detect and measure the mutual relationship. From the viewpoint of neuronal computations, different neuronal morphology may be developed to implement the basic synaptic and somatic operations and thus the scheme to measure the mutual relationship between the past experiences stored in $w_a(t)$ and the neuronal input vector $x_a(t)$. Two specific forms of neuronal morphology will now be discussed. An extention to the morphology of a fuzzy neuron will be addressed in Section 8.3.

8.2.2 Product-Summation-Monotonicity (PSM) Neuronal Form

The *product-summation-monotonicity* (*PSM*) is a widely-used neuronal form in normal neural literature.[14] In order to account for the thresholding operation, let us use the

Foundations of Fuzzy Neural Computations

augmented vector of neural inputs: $x_a(t) = (x_0(t)x_1(t),\ldots,x_n(t))^T \in R^n$, $x_0(t) = 1$, and an augmented vector of synaptic weights: $w_a(t) = (w_0(t)w_1(t),\ldots,w_n(t))^T \in R^n$, defined in (8.5a) and (8.5b), respectively. Using the *PSM* neuronal form, the neuronal processing operations defined in (1-4) can be specialized as

1) Synaptic product:

$$u_i(t) = x_i(t) \cdot w_i(t), \quad i = 0,1,\ldots,n; \tag{8.7}$$

2) Somatic operations:

(i) Aggregation:

$$z(t) = \sum_{i=0}^{n} u_i(t), \quad z(t) \in R, \tag{8.8}$$

(ii) Monotonic nonlinear activation:

$$y(t) = f_m[z(t)], \quad y(t) \in R \tag{8.9}$$

where $x_0(t) = 1$, $w_0(t)$ $(=\theta(t))$ is the value of the neuronal threshold and $f_m[\cdot]$ is a monotonic nonlinear activation function. Theoretically, the only constraints on $f_m[\cdot]$ are monotonicity and boundedness. In practice, $f_m[\cdot]$ is usually a sigmoidal function, a monotonically increasing function. Figure 8.3 gives three kinds of commonly-used monotonic neuronal nonlinear activation functions. Of course, other nonlinear monotonic functions can be employed for different applications. In this form neuro-computational operators consist of "multiplication (synaptic operation)", "summation" (somatic aggregation) together with the somatic thresholding in the ordinary arithmetic. The process of confluence operation (synaptic and aggregation operations in (8.7) and (8.8)) can be defined using an inner product operation of the two vectors, $x_a(t)$ and $w_a(t)$, as shown in Figure 8.4a, Thus, the similarity measure $z(t)$ is given by the confluence operation:

$$z(t) = x_a(t) \odot w_a(t) = x_a^T(t) \cdot w_a(t) \tag{8.10}$$

This process of vector inner product cooresponds to the confluence operation defined in (8.6a-8.6b). This process of confluence operation actually extracts common components (i.e., similarity) between the two vectors: neuronal inputs $x_a(t)$ and the knowledge stored in the neuronal synaptic weights $w_a(t)$, as shown in Figure 8.4a.

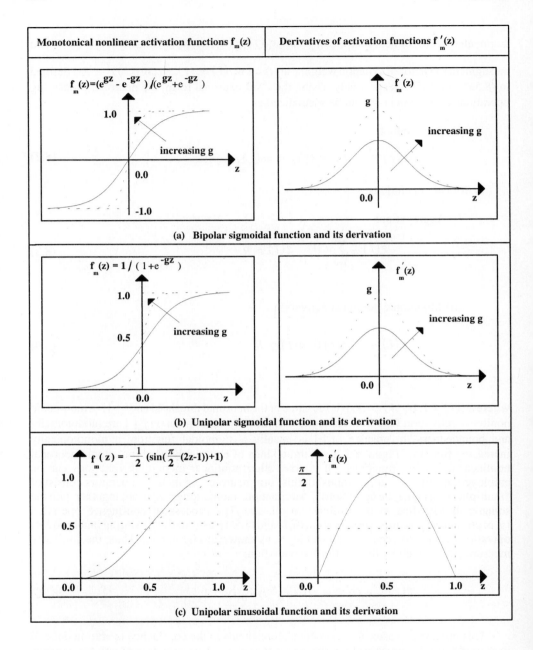

Figure 8.3. Three kinds of possible nonlinear activation functions for *PSM* neuronal form

Foundations of Fuzzy Neural Computations

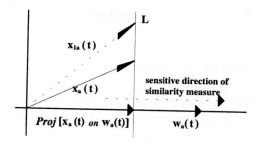

(a) The inner product as a similarity measure, where $Proj[x_a(t) \; on \; w_a(t)]$ denotes the projection of $x_a(t)$ onto $w_a(t)$, L is the line of equal-projection for each $x_a(t)$ with respect to $w_a(t)$

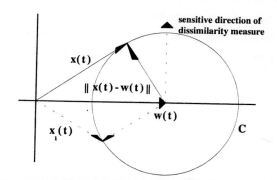

(b) The Eulidean distance as a dissimiliarity measure, where '$\|x(t)-w(t)\|$' denotes the distance between $x(t)$ and $w(t)$, C is the circle of equal-distance for each $x(t)$ with respect to $w(t)$

Figure 8.4. Two kinds of schemes for measuring the mutual relationship between vectors of neuronal inputs and synaptic weights.

This inner product can also be described in terms of vector projection. Assume that $Proj[x_a(t) \; on \; w_a(t)]$ is the projection of $x_a(t)$ onto $w_a(t)$, or alternatively $Proj[w_a(t) \; on \; x_a(t)]$ is the projection of $w_a(t)$ onto $x_a(t)$. When the vector $x_a(t)$ is in the first or fourth quadrant with respect to $w_a(t)$, their inner product $x_a^T(t) \times w_a(t)$ (i.e., mutual relationship measure $z(t)$) will be positive, and will contribute to the excitation of the neuron. Conversely, when the vector $x_a(t)$ is in the second or third quadrant with respect to $w_a(t)$, the inner product $z(t)$ will be negative, and will contribute to the inhibition of the neuron. When both of them are orthogonal to each other, the neuronal activation $z(t)$ will be zero, and the state of the neuron will be unchanged. The purpose of the final sigmoidal nonlinear processing defined in (8.9) is to increase the robustness and flexibility of the neuronal computation. For example, for a bipolar sigmoidal function, for a small $|z(t)|$, the output is almost a linear function of the similarity (mutual relationship) measure $z(t)$. For a large $|z(t)|$, it provides a saturated output bounded by ± 1.

8.2.3 Difference-Summation-Radial (DSR) Neuronal Form

The Gaussian neural networks employs the *difference-summation-radial* (*DSR*) scheme to detect and measure the similarity (mutual relationship) between the fresh neuronal inputs and the past knowledge stored in $w(t)$ of a neuron.[15] In this neuronal morphology, the dissimilarity is first measured by the Eulidean distance between the neuronal input vector $x(t)$ and the synaptic weighting vector $w(t)$. Since the distance measure is always positive, the thresholding cannot be combined with the synaptic and aggregation operations into an integrated mathematical notion, and it should be defined separately. Therefore, let $x(t) = [x_1(t), x_2(t), ..., x_n(t)]^T \in R^n$ as an n-dimensional neuronal input vector; $w(t) = [w_1(t), w_2(t), ..., w_n(t)]^T \in R^n$ as the strength of an n-dimensional synaptic weighting vector. The neuronal operations for this form can be defined as:

1) Synaptic difference:

$$u_i(t) = x_i(t) - w_i(t), \qquad i = 1, 2, ..., n \tag{8.11}$$

2) Somatic operations:
    (i) Aggregation:

$$v(t) = \sum_{i=1}^{n} (u_i(t))^2, \qquad v(t) \in R \tag{8.12}$$

(ii) Thresholding:

$$z(t) = v(t) \oslash \theta(t), \qquad z(t) \in R \tag{8.13}$$

(ii) Radial nonlinear activation:

$$y(t) = f_r(z(t)), \quad y(t) \in [0, 1] \tag{8.14}$$

where $\theta(t) \in R$ is the threshold (bias), '$\oslash$' is a negative-stop subtraction operation, that is, $z(t)$ will be set to zero when $v(t)$ is less than $\theta(t)$. The nonlinear activation function $f_r(\cdot)$ yields an output which is maximum for zero difference (largest similiarity) and decreases for large differences. Since the *DSR* form acturally performs a nonlinear radial basis mapping from $x(t) \in R^n$ to $y(t) \in [0,1]$, the nonlinear activation function $f_r(\cdot)$ should be able to translate into some kind of radial basis functions. Theoretically, one may derive many mathematical forms of theradial basis functions. A common selection for $f_r(\cdot)$ is the exponential function which translates into the commonly-used Gaussian radial basis function, as illustrated in Figure 8.5.

In this form, (8.11-8.13) perform the confluence operation to extract the differential (i.e., mutual dissimiliarity) between the neuronal input vector $x(t)$ and the knowledge stored in the synaptic weighting vector $w(t)$ by the Eulidean distance using a subtraction (synaptic difference) operation, (8.11), a squared summation (somatic aggregation) operation, (8.12), and a bias reduction, (8.13). Finally, (8.14) performs a radial nonlinear activation to give a graded similarity of information. Figure 8.6 gives three kinds of commonly-used nonlinear activation functions employed in the *DSR* neuronal form.

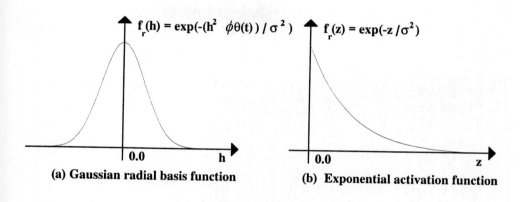

Figure 8.5. The translation of a nonlinear activation function into a radial basis function for the *DSR* neuronal form, where $h^2 = \sum_{i=1}^{n}(x_i - w_i)^2$, $z = h^2 \varnothing \theta(t)$.

This form employs neural computational operations of "subtraction (differential)" and "squared summation (aggregation)" using the ordinary arithmetic. Recalling that the neuronal activation values (i.e., the similarity measure) in the *PSM* neuronal form are produced by the inner product of $x_a(t)$ and $w_a(t)$. This inner product depends on the projection of $x_a(t)$ onto $w_a(t)$ (or $w_a(t)$ onto $x_a(t)$) as shown in Figure 8.4a. In a different way, the neuronal activation values in the *DSR* neuronal form depends on the magnitude of the vector differential between $x(t)$ and $w(t)$ using a second-order Eulidean distance as shown in Figure 8.4b. If the two vectors are in opposite directions to some extent the magnitude of vector difference will be large, and the similarity of information between them will be small. It is only when $x(t)$ and $w(t)$ are equal in magnitude and have the same direction, that the difference is zero and the similarity reaches the maximum.

8.2.4 Some Observations on Two Schemes of Neuronal Computations

Some observations will now be made on the two kinds of neuronal schemes studied in subsections 8.2 and 8.3, the *PSM* and the *DSR* neuronal forms, from computational points of view.

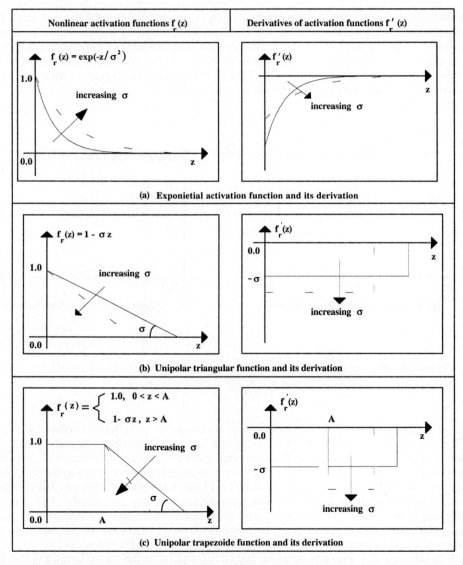

Figure 8.6. Three kinds of possible nonlinear activation functions for *DSR* neuronal form.

In order to determine the similarity of the information between the neuronal inputs $x(t)$ and the synaptic weights $w(t)$, one can either extract the common components directly from $x(t)$ and $w(t)$, or alternatively, their differences can be first extracted and then evaluate the differential measure evaluated. The smaller the differential is, the larger the similarity. In the *PSM* neuronal form, the similarity between $x_a(t)$ and $w_a(t)$ was directly taken as an activation value to the neuron, while the *DSR* neuronal form employs an alternative mechanism using the notion of receptive field. The degree of dissimiliarity is considered as the activation value, but the same destination can be reached in some sense because they employ different types of nonlinear grading functions.

When the sigmoidal function is employed, $f_m[z(t)] = tanh(g \cdot z(t))$, the monotonicity of $f_m[\cdot]$, which is determined by the activation gain g, implies that increasing the activation value (similarity measure) $z(t)$ (more common components) increases the neuronal output. A smaller gain g can relax the activation values, and will produce a *softer* neuronal output, while a higher gain value will produce a *crisper* output. From the neuronal learning perspective, a small gain will allow synaptic weights to adapt quickly and easily to external changes.

In the *DSR* neuronal form, an exponential function is employed, $f_m[z(t)] = \exp\left(-\frac{z(t)}{\sigma^2}\right)$, where $z(t) > 0$, and it is seen that this form of neuronal computation is simply that of performing a Gaussian mapping. The radial basis function determines a spherical receptive field in R^n. A large differential $z(t)$ gives a small output $y(t)$ with respect to a given radial ratio s. On the contrary, a small differential (a large similarity between $x(t)$ and $w(t)$) will cause a large output. Here $w(t)$ provides a center to the spherical receptive field in R^n, while the radial ratio s localizes it. That is, the radial ratio s controls the radius of the spherical field. Decreasing s shrinks the receptive field which means that only those neuronal inputs $x(t)$ which are most common to $w(t)$ will activate the neuron to a high degree.

Another observation is on the individual performance of the two schemes on measuring the similarity between $x(t)$ and $w(t)$. Recalling that in the case of similarity measured by the inner product as shown in Figure 8.4a, there is a equal-projection line L for each input vector $x_a(t)$ with respect to $w_a(t)$, $Proj[x_{1a}(t) \, on \, w_a(t)] = Proj[x_a(t) \, on \, w_a(t)]$, that is, $x_{1a}(t)$ has the same amount of similarity with $w_a(t)$ as $x_a(t)$. The sensitive direction of this kind of measuring scheme is along the weighting vector $w_a(t)$. Therefore, when a neural network is employed using this scheme to approach to the functional approximation, what must be done is to adjust the weighting vector $w_a(t)$ so that every point in the functional space to be approximated is sensitive to each other along the direction of $w_a(t)$. On the contrary, in the case of dissimiliarity measured by the Eulidean distance in the *DSR* neuronal form, there is a equal-distance circle C for each input vector $x(t)$ with respect to $w(t)$, $\|x_1(t) - w(t)\| = \|x(t) - w(t)\|$, i.e., $x_1(t)$ which has the same amount of dissimiliarity with $w(t)$ as $x(t)$, Figure 8.4b. In this case the sensitive direction of the measuring scheme is on the radial vector from the end-point of the weighting vector $w(t)$. Therefore, when a neural network is applied using this measuring scheme to solve the problem of pattern classification, the objective of the learning scheme is to adjust $w(t)$ so that there is a different amplitude of the radial with respect to $w(t)$ for each potential patterns (classes). This kind of view may be used to explain the experimental observations given by Hush in[29] that a *Radial* network provides a more efficient solution to a pattern classification problem than a *Sigmoidal* network, while the *Sigmoidal* network usually solves the problem of functional approximation more efficiently than the *Radial* network.

8.3 A GENERALIZED MORPHOLOGY OF FUZZY COMPUTATIONAL NEURONS

In the previous section, two generalized computational neuronal morphologies were given. In this section, this morphology to a fuzzy situation is extended from the viewpoint of neuronal operational logic. When the neuronal input $x(t)$ is an n-dimensional vector of real-valued degrees of fuzzy membership functions; that is, $x(t) \in I^n = [0, 1]^n$, the two forms of neuronal computations which are described in detail in Section 8.2 are still applicable. However, when either $x(t)$ is represented by a vector of fuzzy sets for the observations with linguistic uncertainties, or $w(t)$ is represented by a vector of fuzzy sets for the accumlated knowledge with cognitive uncertainties, an alternative form, named *Fuzzy Operation Form* (*FOF*), should be employed.

The objective of combining neural networks with fuzzy logic is to enhance the information processing capabilities of neuronal computations with cognitive (linguistic) uncertainties. The basic mechanism of information processing in a neuro-processor presented in Section 8.2 is still applicable to a fuzzy neural-processor, but is implemented by using a fuzzy computational logic.

Again, the neuronal computational procedure in a fuzzy neuron includes two kinds of operations:

(i) the *synaptic fuzzy operation,* and
(ii) the *somatic fuzzy operation*.

Some geometrical interpretation on the current theory of fuzzy sets will now be extended.

8.3.1 Geometric Interpretation of Fuzzy Neuronal Operations

The current theory of fuzzy sets is not sufficiently complete to construct the mathematics of fuzzy neuronal models for the neuro-fuzzy approach. For example, difficulties may arise in describing the fuzzy confluence and fuzzy activation for a fuzzy neuron, and in developing the learning and adaptive algorithm using the traditional fuzzy-set theoretic approaches. It appears, therefore, that a new computational logic needs to be developed for fuzzy neuronal computations, called *neuro-fuzzy logic*. Kosko[1] proposed a fuzzy-set-as-point view which gives an alternative and intuitive interpretation to a fuzzy set compared to a crisp set. No further interpretations were studied, however, for various mathematical operations between fuzzy sets using that view. Strictly speaking, this viewpoint has some limitations in its application to the modeling of fuzzy neurons. In fact, this viewpoint is based on the discretization over the range of a fuzzy variable in a fuzzy set.

Starting from the basic biological viewpoint in a fuzzy neuron, some fundamental mathematics are proposed in the following section for its use in fuzzy neuronal operations. A fuzzy vector space and related operations between vectors are defined and described in detail in the Appendix.

8.3.2 Fuzzy Sets as Vectors

Kosko[1] identifies a fuzzy set with a point in a unit hypercube S. This unit space is constructed with the discretized values of the membership function of a fuzzy set as the value of each coordinate of the space. A more natural and intuitive approach, however, is to represent a fuzzy set using a fuzzy vector x as defined in the Appendix, and as shown in Figure 8.7. More importantly, when the concept of a fuzzy set is extended to include the case where each element may come from different universes of discourse, or from different original fuzzy sets in the same universe of discourse, they can be represented by the extended fuzzy sets by *Type One* or *Type Two* fuzzy vectors (see the Appendix). This extension is most useful because a practical fuzzy logic system commonly takes fuzzy data from the different universes of discourse or different fuzzy sets as inputs. Based on this extension, a logical combination between two fuzzy sets will be naturally interpreted into a geometric operation between two fuzzy vectors in a fuzzy space S.

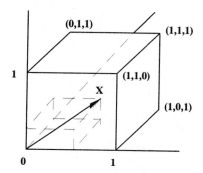

Figure 8.7. Fuzzy unipolar hypercube and fuzzy vectors.

In the following, it is shown that this fuzzy-set-as-a-vector (S, x) defines an intuitive geometric interpretation of finite fuzzy neuronal operations. In this regard, the following are to be noted:

(i) the difference between two fuzzy vectors, x and y, in the same fuzzy space can be measured by the mth-order distance between them, usually defined by $D_m(x,y) = \sqrt[m]{\frac{1}{n}\sum_{i=1}^{n}(x_i - y_i)^m}$;

(ii) when a fuzzy set is interpreted as a fuzzy vector x in S, the cordiality or size of the fuzzy set is equal to the fuzzy Hamming distance between the vector x and the zero element in S, that is, $M(x) = n \times D_1(x, 0)$, which is naturally equal to the magnitude of the vector, where n is the dimension of the vector;

(iii) those vectors which take 2^n vertices of the cube I^n as their endpoints define 2^n nonfuzzy sets, and these 2^n nonfuzzy sets constitute the power set of the ordinary (nonfuzzy) set z;

(iv) the vector which takes the midpoint with its fuzzy elements of the vector equal to 0.5 of the hypercube as its endpoint is maximally fuzzy;

(v) the fuzziness of a fuzzy set represented by a vector can be measured by a geometric fuzzy entropy, $E(x) \in [0, 1]$, defined by $E(x) = a/b = D_m(x, x_{near})/D_m(x, x_{far})$, where x_{near} is the nonfuzzy vertex nearest to x, while x_{far} the nonfuzzy vertex vector farthest from x, as shown in Figure 8.8.

Fuzzy Mapping Operations A mapping from fuzzy sets to fuzzy sets can be considered as a transformation from one fuzzy space to another. The simplest case is that of transforming a fuzzy vector $x \in I^n$ into a fuzzy vector $y \in I^m$. This can be done by fuzzy vector-matrix multiplication defined in (42) in the Appendix. In order to map a fuzzy vector x into

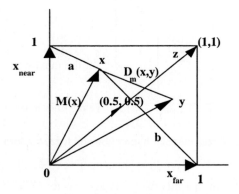

Figure 8.8. Fuzzy sets as vectors.

a fuzzy vector y, a fuzzy matrix M needs to be first constructed using the correlation-product operation defined in (8.43) in the Appendix as is done during a learning phase in ordinary associative memory approaches,[1] or using other adaptive learning schemes,[22] then the vector-matrix multiplication is performed in an operation phase for a fresh input vector. It should be noted that:

(i) a fuzzy matrix M usually stores enough information to map x into y, where M is not unique. This means that different versions of M may be constructed to perform the same mapping similar to the case where different versions can be constructed in the physical implementation of a system to meet its functional requirements in systems engineering;

(ii) the information in M represents the accumulated knowledge acquired to perform the given mapping and should be continuously adapted to external changes in an adaptive fuzzy logic system;

(iii) M could be constructed and represented as fuzzy synaptic weights in a fuzzy neuronal model;

(iv) if M is an n-dimensional row fuzzy vector, then the vector-matrix multiplication degenerates into an fuzzy inner product in I^n defined in (8.39), and y will be a real-valued fuzzy degree. In the context of neuronal operations, the vector-matrix multiplication corresponds to the confluence operation (synaptic operation and somatic aggregation) in a fuzzy neuron;

(v) if m equals to n then the vector-matrix multiplication actually conducts a vector rotation in I^n, or a vector filtering from the viewpoint of information processing.

Fuzzy Vector Projection Another important type of operation between fuzzy vectors in a fuzzy space I^n is the fuzzy vector projection. Generally speaking, the projection of a vector x onto a vector y, denoted by $Proj[x\ on\ y]$, reflects the degree of similarity between x and y. The process of projecting x onto y is equivalent to measuring x in the region covered by (or common with) y. In a real or complex space, the common region is exactly represented by the length and direction of the vector. When the vector x is orthogonal to the vector y there is no similarity between them. When they are equal in magnitude and have the same direction, the common information that is obtained is at the maximum. In a fuzzy vector space, however, the problem is how to measure the similarity between fuzzy vectors x and y ? As discussed in Section 8.3.1.1, the difference between fuzzy vectors x and y may be measured by the mth-order distance. Intuitively, however, this is not exactly the inverse measure index of the similarity between x and y. On the other hand, the definition of the similarity depends on the specific arithmetic operation that is employed in the fuzzy space. As we known, a fuzzy vector x determines its power set $F(2^x)$ at the same time, which is hyper-rectangle snug with side lengths equal to the elements in x. Hence, the power set $F(2^x)$ determines the common region of x. All vectors in the hyper-rectangle are subsets of the vector x, while different vectors with their endpoints outside the hyper-rectangle resemble subsets of x or are common to x with different degrees. An example, $z \in F(2^x)$, is shown in Figure 8.9, which means that the information in z is totally covered by x; that is, the similarity between x and z is maximal. On the other hand, y is not completely covered by x, and the maximal similarity between x and y is v. Hence, the projection of x onto y is intuitively defined to be v.

The projection is also a fuzzy vector in the fuzzy vector space. Based on the fuzzy *MIN-MAX* operations, the projection v can be defined as

$$v = Proj[x\ on\ y] = (MIN\ \{x_i, y_i\}\ |\ i = 1, 2, ..., n)^T \qquad (8.15)$$

An interesting case is that the projection of y onto x is also v. This symmetric feature mainly results from the lattice structure of the fuzzy vector operation in terms of the *MIN-MAX* operators. Hence, if x is orthogonal to y, then $Proj[x\ on\ y] = 0$. This only happens when both x and y degenerate to fuzzy vectors with some elements (coordinate values) in x and y

to zeros. When $x = y$, $Proj[x \ on \ y] = y = x$.

Based on this definition it is seen that the fuzzy inner product of x and y defined in (8.39) can be rewritten as

$$x \ (\cdot) \ y = x \ (\cdot) \ Proj[y \ on \ x] = y \ (\cdot) \ Proj[x \ on \ y] \tag{8.16}$$

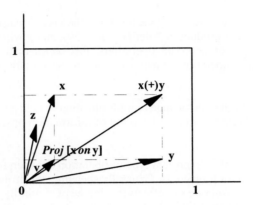

Figure 8.9. Geometric interpretation of projection of fuzzy vectors.

Consider now a more intuitive explanation of (8.16). Assuming that there are two vectors: $x = (0.3, 0.7)^T$ and $y = (0.8, 0.2)^T$, defined in I^2. Their inner product for a crisp operation can be obtained as

$$x^T \cdot y = ((0.3 \times 0.8)+(0.7 \times 0.2)) = 0.38$$

where '$\times$' and '+' are operators of *multiplication* and *summation*, respectively, defined in the crisp arithmetic.

This operation can be extended to fuzzy situation. The fuzzy inner product can intuitively be extended by employing *MIN-MAX* fuzzy operations by replacing the multiplication operator and the summation operator, respectively, as:

$$x \ (\cdot) \ y = MAX\{\ MIN\{0.3, 0.8\}, \ MIN\{0.7, 0.2\}\} = 0.3$$

On the other hand, according to (8.15), the projection of x onto y is $v = (0.3, 0.2)^T$. Therefore, their fuzzy inner product can also be obtained by (8.16):

$$x \ (\cdot) \ y = (0.8, 0.2)^T (\cdot) \ (0.3, 0.2)^T = MAX\{\ MIN\{0.8, 0.3\}, \ MIN\{0.2, 0.2\}\} = 0.3$$

8.3.2 Synaptic Fuzzy Operation

Based on the basic operations and features of the fuzzy vectors in the hypercube described above, the fuzzy neuronal operations can now be defined. Note that the thresholding operation in the fuzzy neuronal computations usually needs to be defined separately. Therefore, we take an n-dimensional vector $x(t) = (x_1(t), x_2(t), ..., x_n(t))^T \in \Re^n$ as the neuronal fuzzy inputs, and an n-dimensional vector $w(t) = (w_1(t), w_2(t), ..., w_n(t))^T \in \Re^n$ as the synaptic fuzzy weights.*

The synapse in a fuzzy neuron is considered as a storage element of the past experience (knowledge). It learns from the neuronal environment and continuously adapts its strength. This knowledge appears in the form of synaptic fuzzy weights, where a synaptic fuzzy weight may be a real-valued degree of a fuzzy membership function over the unit interval [0, 1], or a fuzzy set represented by its membership function, which depends on the notion of fuzzy neuronal inputs. The synaptic fuzzy operation provides a fuzzy measure of mutual relationship between the past experience stored (memorized) in the synapse represented by a fuzzy weight vector $w(t)$ and a fuzzy input vector $x(t)$ at the time t given by

$$u_i(t) = w_i(t) \ominus x_i(t), \quad i = 1, 2, ..., n, \quad u_i(t) \in \Re \quad (8.17)$$

where '$\ominus$' represents the synaptic fuzzy operation. If $x(t)$ is an n-dimensional fuzzy vector bounded by a membership function over the unit interval [0, 1], $w(t)$ is also an n-dimensional fuzzy weight vector represented by a graded membership over the unit interval [0, 1]. If $x(t)$ is represented by a vector of fuzzy sets, $w(t)$ is also represented by a vector of synaptic weighting fuzzy sets.

8.3.3 Somatic Fuzzy Operation

The synapse transmits the results to the soma after the synaptic fuzzy operation defined in (8.17). Somatic fuzzy operations are then performed on them through a two-step process:

(i) *fuzzy aggregation,* and
(ii) *nonlinear fuzzy activation with thresholding.*

Fuzzy Aggregation The fuzzy aggregation operation, the first step of somatic operation, is actually a fuzzy mapping from $u(t) \in \Re^n$ into a real-valued degree of membership function $v(t) \in \Re$, or into a single fuzzy set represented by its membership function.

$$v(t) = \oiint (u_i(t) \mid i = 1, 2, ..., n), \quad v(t) \in \Re \quad (8.18)$$

where '$\oiint$' denotes fuzzy aggregation. Using the definition of fuzzy mapping described before, (8.18) may be rewritten as:

$$v(t) = M(t) \, (\mathrm{o}) \, u(t) \quad (8.19)$$

where $M(t)$ is a $1 \times n$ row unit vector, and '(o)' is the fuzzy vector-matrix multiplication

---

*$\Re$ is either a unipolar unit interval [0, 1], or a bipolar unit interval [-1, 1], or a real number R. Thus, $\Re^n$ is an n-dimensional hypercube in $[0, 1]^n$, or $[-1, 1]^n$, or an n-dimensional space R^n.

defined in (8.42). In terms of fuzzy inner product, (8.17) and (8.18) may be combined into:

$$v(t) = w(t) \, (\cdot) \, x(t) \tag{8.20}$$

where '(·)' is the fuzzy inner product.

Thresholding and Nonlinear Fuzzy Activation It is the nonlinear operation in the soma of a biological neuron that provides the neuronal process with intriguing behavioral properties. For a fuzzy signal (e.g., a fuzzy set), the fuzzy thresholding and nonlinear activation can be defined as follows:

Thresholding: $\qquad z(t) = v(t) \, (\text{-}) \, \theta(t), \qquad z(t) \in \Re \tag{8.21}$

Nonlinear activation: $\qquad y(t) = f[z(t)], \qquad y(t) \in \Re \tag{8.22}$

where "(-)" is a fuzzy thresholding and $f[\cdot]$ is a nonlinear fuzzy activation function. Some specific cases of the generalized fuzzy neuronal model will be discussed in Section 8.4.

From the above discussion, it is clear that the fuzzy operators: 'Θ', '$\#$', and '(-)' defined in (8.17), (8.18) and (8.21) serve as an integrated operation: *fuzzy confluence,* between the neuronal fuzzy inputs and the past experience stored in the synaptic fuzzy weights.* Together with the nonlinear fuzzy activation function, they define a kind of computational logic for fuzzy neuronal operations.

Following the neuronal operations developed in Section 8.2, two kinds of schemes can be implemented for measuring similarity of information by using different combinations of computational operators in the fuzzy arithmetic. For example, the fuzzy subtraction defined in (8.37) in the Appendix can be employed for the synaptic fuzzy operation, the fuzzy addition defined in (8.36) for the neuronal fuzzy aggregation operation, and again the fuzzy subtraction for rhe fuzzy thresholding operation. The activation function can be defined by $f[\cdot]$, a monotonic nonlinear function such as the sinusoid function over [0, 1] as shown in Figure 8.10a. In other words, the fuzzy inner product defined in (8.20) can be employed to directly extract the similarity information between $x(t)$ and $w(t)$.

In addition, for $x(t) \in I^n$, $w(t) \in I^n$, by using the definitions of fuzzy operations over a unit hypercube, an alternative combination of computational operators for the second measuring scheme of mutual relationship, the *DSR*, can be described as

$$u_i(t) = x_i(t) - w_i(t), \quad i = 1, 2, ..., n, \quad u_i(t) \in [-1, 1]^n \tag{8.23}$$

$$v(t) = \sqrt{\frac{1}{n} \sum_{i=1}^{n} (u_i(t))^2}, \quad v(t) \in I \tag{8.24}$$

$$z(t) = v(t) \oslash q(t), \quad z(t) \in I, \, \theta(t) \in [-1, 1] \tag{8.25}$$

---

* The past experience stored in the synaptic weights (including the threshold) may be termed as a long term memory (LTM).

Foundations of Fuzzy Neural Computations

$$y(t) = f_r[z(t)], \quad y(t) \in I \tag{8.26}$$

where (8.23) and (8.24) calculate the differential between $x(t)$ and $w(t)$ using the second-order distance defined in a fuzzy space, (8.25) performs a bias reduction to the differential $v(t)$, where 'Ø' means a negative-stop subtraction operation, that is, 'Ø' serves as the normal subtraction operator, but $z(t)$ will be zero when $v(t)$ is less than $\theta(t)$. In (8.26) $f_r[\cdot]$ is a nonlinear activation function over I and which translates the neuronal operation into a radial basis mapping. One such a function is shown in Figure 8.10b, where $f_r[\cdot] = \exp[-(z(t))/((1-z(t))\cdot\sigma^2)]$ over the unipolar interval [0, 1] for $z(t) \in [0, 1]$, and s is the radial ratio associated with the fuzzy neuron.

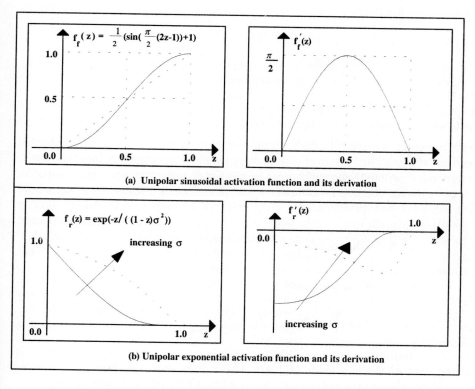

Figure 8.10. Two kinds of nonlinear activation functions for fuzzy neuronal form.

8.4 COMPUTATIONAL FUZZY NEURAL NETWORKS

The most important objective of integrating fuzzy logic systems with neural networks is to deal with the fuzzy information in a more flexible and adaptive way. Studies on this integration have been conducted in many aspects of research from learning algorithms to implementation of architecture, and from approximation theory to cognitive analysis. Several different methods of classifying current approaches or models of fuzzy neurons may be available. Gupta[2] gave a general review that categorized these models into non-fuzzy neuronal models and fuzzy neuronal models. In the following subsections, some further discussions on neuronal models are presented.

8.4.1 Non-Fuzzy Neuronal Models (FNN_I)

A standard non-fuzzy neuronal model, denoted by FNN_I, is usually a feedforward layered neural network. In this FNN_I, the input vector $x(t)$ can be used the fuzzy information, usually represented by an n-dimensional vector of real-valued membership grades; that is, $x(t) \in [0, 1]^n$. If it is assumed that the output $y(t)$ of the network is an m-dimensional vector, $y(t) \in [0, 1]^m$, this multi-layered neural network actually performs a nonlinear mapping from an n-dimensional hypercube $I^n = [0, 1]^n$ to an m-dimensional hypercube $I^m = [0, 1]^m$, that is,

$$x(t) \in [0, 1]^n \xrightarrow{FNN_I} y(t) \in [0, 1]^m$$

The weights in this network may be real numbers; that is, $w(t) \in [-\infty, \infty]$. The learning algorithm for this kind of network can be any appropriate learning algorithm used in the standard neural networks, such as the backpropagation algorithm.

In practical applications, an element $x_i(t)$ ($i = 1, 2, ..., n$) in $x(t)$ usually represents the degree to which an input fuzzy variable $\bar{x}_i(t)$ belongs to a fuzzy set, while $y_j(t)$ ($j = 1, 2, ..., m$) in $y(t)$ represents a degree of a fuzzy output variable $\bar{y}_j(t)$ in a fuzzy set. This approach has been intuitively applied in many fields. In fuzzy pattern recognition[8], the input variable $\bar{x}_i(t)$ corresponds to an input feature which can be a linguistic variable or a numerical variable. In both cases the membership function of the feature variable in a feature fuzzy set should be first extracted and quantized. The crisp value of an output node usually directly corresponds to the grade to which the input pattern is ranked into a known class associated with the node. The final decision can then be made based on the output vector of graduation. Pal[8] applied this method with a π quantization for the membership function of the input sensor-based feature signals into a speech recognition.

In the areas of fuzzy decision-making, aggregation functions for both the evidence aggregation and the hypothesis aggregation exhibit intrinsic features of fuzziness, nonlinearity and adaptation in realistic cases. A promising approach to these aggregations is to develop a FNN_I model to perform a nonlinear and adaptive mapping from an evidence space or a hypotheses space into a consensus space.[16,17]

In the diagnostic domain, when the diagnostic reasoning with fuzzy uncertainties is considered, one of the two kinds of fuzzy associations should at least be available: experiential associations (from symptoms to disorders) or causal associations (from disorders to symptoms). Artificial intelligence approaches employ the extended deductive reasoning method with experiential associations (e.g., well-known fuzzy modus ponens)[18] and the abductive reasoning method with causal associations (e.g., fuzzy explanatory model).[19] Human diagnostic experts, however, would not rather evaluate the above associations in a simple fuzzy relation matrix (or rules) because the mappings are commonly too complex (multiple to multiple, nonlinear, adaptively changing), especially in the diagnosis of multiple disorders. Using the fuzzy neural approach to this problem leads to the easy construction of a very effective FNN_1 model having symptoms (grades of presence) as its inputs and disorders (occurring degrees) as its outputs.[7] However, it is not easy to implement an abductive reasoning using a straightforward FNN_1 model. In this case, apart from the causal mapping, the competition between the two sets of disorders for explaining a given set of symptoms should be taken into account.[17,20]

In the adaptation of fuzzy logic systems, the membership functions are generally supposed to be *given*, or they are assumed to be triangular, trapezoid or Gaussian shapes. In practice, some empirical studies on pointwise measurements $(x_i, \mu_A(x_i) | i = 1, 2, ..., N)$ are not too difficult to be done. An intuitive problem is how to model and adapt the membership function using the given measuring points. In order to solve this kind of problem a FNN_1 model can be constructed with the given points as training samples. In this case the input to the model is the universe of discourse of a fuzzy set the fuzzy variable belongs to, while the output is the membership function approximation. This process corresponds to an adptively nonlinear regression.

Furthermore, non-fuzzy neuronal networks can also be applied to model fuzzy systems and to the design of neuro-fuzzy controllers.[9]

These non-fuzzy neuronal networks have the advantages of simplicity and readily available learning algorithms. For some specific problems they are effective. However, this approach to fuzzy information processing suffers from the shortcomings of the "*black box*" approach. What the network recalls, what it can not recall, and what inferential paths it takes are not known. That is why difficulties are experienced in explaining the rationality of results obtained by the network. These drawback sometimes prevent studies sometime of the effective combination of fuzzy logic systems and neural networks.

8.4.2 Fuzzy Neuronal Models

From the viewpoint of neuronal computation for information processing, fuzzy neuronal models can be classified into two categories. One is a category for the processing of quantized fuzzy information, having two vectors of real-valued membership degrees as its inputs and outputs, respectively. It is called the fuzzy-valued neural networks (FNN_2). The other is a category employed to directly deal with the continuous fuzzy sets, taking two vectors of fuzzy sets as its inputs and outputs, respectively. It is called the fuzzy-sets neural networks (FNN_3).

Fuzzy-valued Neural Networks (FNN_2) A standard fuzzy valued neural model takes a real-valued vector of membership degrees of input fuzzy variables as its inputs; that is, $x(t) \in [0, 1]^n$, and a real-valued vector of membership functions of output fuzzy variables as its outputs; that is, $y(t) \in [0, 1]^m$. In most situations $n > m$. All neuronal operations in this model are taken by the appropriate fuzzy operators. This type of fuzzy neurons can approximate fuzzy mapping functions in a quantized form. A mathematical morphology for FNN_2 can be easily derived from the standard fuzzy operational form of the neurons discussed in Section 8.3. Some particular forms of the model will now be discussed, as well as some problems associated with the current FNN_2 approach. A new model of FNN_2 is then proposed.

Inclusions and Limits of FNN_2 Briefly speaking, if the synaptic fuzzy operation in (8.17) is replaced by the fuzzy MIN-operation; that is, $u_i(t) = MIN\{w_i(t), x_i(t)\}, i = 1, 2, ..., n$, while the fuzzy aggregation employs the fuzzy *MAX*-operation, that is, $v(t) = MAX\{u_i(t) | i = 1, 2, ..., n\}$, then the neuronal model is simply the widely-used *MIN-MAX* version of the fuzzy neurons.[22, 23] However, if the synaptic fuzzy operation is replaced by the ordinary arithmetic product; that is, $u_i(t) = w_i(t) \cdot x_i(t)$, $i = 1, 2, ..., n$, while the fuzzy aggregation still employs the fuzzy *MAX*-operation, then the neuronal model is the same as that discussed in.[6] If the synaptic fuzzy operation is replaced by the arithmetic product, and (8.19) is replaced by $v(t) = M^T(t) \cdot u(t)$, where $M(t)$ is an ordered weighting vector, the neuron is simply an *OWA* neuron as proposed by Yager,[21, 24] except that there is no nonlinear activation in the *OWA* neuron. Some other *ad hoc* forms of fuzzy neurons can also be specially derived from the generalized neuronal model presented in Section 8.3.

Current studies on FNN_2 suffer from several problems. They include the fact that the widely-used *MIN-MAX* operation is either very pessimistic or optimistic. In many cases, fuzzy neural models with such neurons are only sensitive to the inputs from some neuronal nodes. This may cause the loss of the feature of massive inter-connections which is one of most important properties in neural computation because this architecture is only sensitive to some particular neurons. This problem can be corrected to some extent by replacing the *MIN-MAX* operators with some other t-norms and t-conorms. Another problem is associated with the lattice structure. It has been shown that the *MIN-MAX* lattice operation cannot cope with the main essence residing in repetitive experiments.[25] The *AND*-like operation expressed by the minimum favours the minimal element even though other elements have large grades of membership. The opposite situation happens in the case of *OR*-like operation expressed by the maximum. Another drawback occurring within the lattice structure of fuzzy sets is the lack of a convenient way of dealing with *negative* information. Since nothing is explicitly said about the degree to which an element $x_i(t)$ does not belong to a set A, by default it could be stated that the degree of belongingness of $x_i(t)$ to the complement of A should increase with the decrease of $\mu_A(x_i)$.[25] However, the lack of explicit expression of negative information may result in some difficulties in fuzzy neuronal computations. For example, if it is assumed that a positive number is an excitation to a neuron, then there is no active inhibitive representation in such a fuzzy neuron. Even though those elements which are less than 0.5 are taken as some kind of inhibitive information because of the non-decreasing features of the fuzzy *MAX*-operation, no inhibitive information will actually contribute to the final activation in (8.20).

An Extension to Bipolar Interval of Membership Function For these reasons it is recommended[11] that the unipolar unit interval [0, 1] of membership function of input fuzzy vector $x^*(t)$ is extended to a bipolar interval [-1, 1], using an instant transformation:

$$x^*(t) \in [0, 1]^n \longrightarrow x(t) \in [-1, 1]^n \quad \text{such that } x(t) = 2x^*(t) - 1$$

where -1 stands for the complete exclusion, and +1 expresses a degree of complete membership.

The greater is the absolute value of the membership function, the more certain is the information about the object, where 0 refers to the case where the element is neither accepted to the class nor excluded from it. The higher is the absolute value of the grade of membership, the smaller is the uncertainty allocated to the corresponding element of the universe of discourse. For a fuzzy neuron, a positive activation value will excite it while a negative activation value will inhibit it. In order to process such a symmetrical information some basic operations are defined in the bipolar hypercube space $II \in [-1, 1]^n$. Assume that $x, y \in [-1, 1]^n$, $x_i, y_i \in [-1, 1]$, $M_{n \times n}$ is fuzzy matrix

Foundations of Fuzzy Neural Computations

defined in $[-1, 1]^{n \times n}$, then we define the following operations:

(i) Addition:
$$x_i (+) y_i = \frac{1}{2}(x_i + y_i)$$

(ii) Subtraction:
$$x_i (-) y_i = \frac{1}{2}(x_i - y_i)$$

(iii) Multiplication:
$$x_i \otimes y_i = x_i \times y_i$$
or
$$\lambda \otimes x_i = \text{sign}(\lambda \cdot x_i) \, MIN \, \{1, |\lambda \cdot x_i|\}, \, \lambda \in [-\infty, \infty]$$

(iv) Inner product:
$$x \, (\cdot) \, y = \left(\frac{1}{n} \sum_{i=1}^{n} ((x_i \cdot y_i) \, \bigg| \, i = 1, 2, \ldots, n \right)^T$$

(v) Vector-matrix multiplication:
$$x = M \, (\text{o}) \, y = ((M_i(\cdot) \, y) \, | \, i = 1, 2, \ldots, n)^T$$
where M_i is the ith row vector of M.

(vi) Correlation product:
$$M = x \, \copyright \, y = ((x_i \cdot y_j) \, | \, i = 1, 2, \ldots, n; \, j = 1, 2, \ldots, n)^T$$

A New Neuronal Model: FNN$_2$ Based on the basic operations defined above, a new fuzzy-valued computational neuronal model can be derived. Assume that an input fuzzy vector $x(t) \in [-1, 1]^n$ is put into such a fuzzy neuron. Then the following neuronal operations can be defined:

Synaptic operation: $\quad u_i(t) = w_i(t) \otimes x_i(t), \quad u_i(t) \in [-1, 1]$ (8.27)

Aggregation operation: $\quad v(t) = (+) \, (\, u_i(t) \, | \, i = 1, 2, \ldots, n)$ (8.28) where $w_i(t) \in [-1, 1]$ is a synaptic fuzzy weight. Combine (8.27) and (8.28), and defining the confluence operation for similarity measure $v(t)$ as:

Confluence operation:

$$v(t) = w(t) \, (\cdot) \, x(t), \quad v(t) \in [-1, 1] \tag{8.29}$$

the somatic nonlinear activation operation is:

Nonlinear activation:

$$y(t) = f[v(t)], \quad y(t) \in [-1, 1] \tag{8.30}$$

where $f[\cdot]$ is a bounded nonlinear activation function. In this presentation the bounded monotonic sinusoid function is employed, that is:

$$y(t) = sin\left(\frac{\pi}{2} v(t)\right) \tag{8.31}$$

Learning Algorithm for The Proposed Model The backpropagation mechanism will now be derived for the proposed FNN_2 model for the supervised learning with the synaptic weights $w(t)$. The process of adaptation of the synaptic weights may be described as

$$w(t+1) = f[w(t), \Delta w(t)] \tag{8.32}$$

for the discrete case, where $f(\cdot)$ is an appropriate form of the adaptive function. The change of $w(t)$, $\Delta w(t)$, is determined by using the gradient descent method, as

$$w(t) = -\eta \left(\frac{fE(t)}{fw_i(t)} \bigg| i = 1, 2, \ldots, n \right)^T \tag{8.33}$$

In the above, η is the learning parameter, t is the number of iteration, and $E(t)$ is the output error function defined by

$$E(t) = \sum_p E_p(t) = \sum_p \left(d_p - y_p(t)\right)^2 \tag{8.34}$$

where d_p is the desired output, and $y_p(t)$ is the neuronal output at the t-th iteration for the p-th sample. Using the chain rule

Foundations of Fuzzy Neural Computations

$$\frac{fE(t)}{fw_i(t)} = \frac{1}{2}\sum_p \frac{fE_p(t)}{fw_i(t)}$$

$$= \sum_p (d_p - y_p(t))\frac{fy_p(t)}{fw_i(t)}$$

Then, by (8.29)

$$\frac{fy_p(t)}{fw_i(t)} = f[v_p(t)]\frac{fv_p(t)}{fw_i(t)}$$

$$= -\frac{\pi}{2n}\cos\left(\frac{\pi}{2}v_p(t)\right)\cdot x_{p,i}$$

where $f[\cdot]$ is defined in (30-31). Further, let $e_p(t) = d_p - y_p(t)$, then

$$\frac{fE(t)}{fw_i(t)} = \sum_p \left(\frac{p}{2n}\cos(v_p(t))\cdot e_p(t)\cdot x_{p,i}\right) \quad (8.35)$$

Combining (8.32-33) and (8.35), a generalized delta rule for fuzzy neuronal learning can be obtained. Obviously, this backpropagation mechanism can easily be generalized to the case for multi-stage fuzzy-valued neurons.

Fuzzy-Sets Neural Networks (FNN$_3$) For a neural network to be called a *FNN$_3$*, the inputs to the network, the weights in the network, and the outputs from the networks must be fuzzy sets as shown in Figure 8.11. These fuzzy sets are represented by their whole membership functions. In the following a bar is placed over a uppercase symbol if it denotes a fuzzy set, while the lowercase letters represent real numbers. Thus $\overline{O}, \overline{W}, \overline{Z}, \overline{U}, \overline{V}, \ldots$ are all the fuzzy sets. For these neurons, the continuous forms of the fuzzy sets do not need to be quantized. Unfortunately, little progress has been made so far on the design (learning) of such *FNN$_3$*, because of their high complexity. Consider the *i*-th neuron in the *k*-th layer in Figure 8.11, where

$$\overline{U}_{i,j,k} = \overline{W}_{i,j,k} \Theta \overline{O}_{j,k-1}$$
$$\overline{V}_{i,k} = \oiint(\overline{U}_{i,j,k}|j=1,2,\ldots,N_{k-1})$$
$$\overline{Z}_{i,k} = \overline{V}_{i,k}(-)\overline{\theta}_{i,k}$$
$$\overline{O}_{i,k} = f[\overline{Z}_{i,k}]$$

$$i = 1, 2, ..., N_k, \quad j = 1, 2, ..., N_{k-1}, k = 1, 2, ..., L$$

where $\overline{O}_{j,k-1}$ is a fuzzy set of output of the j-th neuron in the (k-1)th layer; $\overline{W}_{i,j,k}$ is the fuzzy strength of connections between the i-th neuron in the k-th layer and the j-th neuron in the (k-l)th layer; $\overline{\theta}_{i,k}$ is a fuzzy threshold associated with the i-th neuron; $f[\cdot]$ is the nonlinear activation function for a fuzzy-sets neuron, 'Θ', '$\#$', '(-)' are fuzzy operators in the neuron for 'synaptic operation', 'aggregation operation', 'thresholding operation', respectively, and L is the number of layers in the FNN_3, N_k is the number of neurons in the k-th layer.

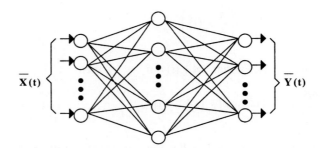

Figure 8.11. Feedforward fuzzy-sets neural networks.

Even without interconnection between the neurons within a layer as shown in Figure 8.11; that is, a layered feedforward fuzzy-sets neural networks, there is no effective learning algorithm for such a network if there is no prior knowledge about the fuzzy sets and operators between them. What exists in current literature on fuzzy neural networks is only the generalized model. In order to simplify the problem, many researchers have assigned specific shapes of membership functions to these fuzzy sets as the prior knowledge.[4, 27]

It should be noted that there is an aberrant form of FNN_3; that is, $\overline{W}(t)_{i,j,k}$ degenerates to a real number $w_{i,j,k}$, but that the initial inputs $x(t)$ to the network are still fuzzy sets. Another aberrant form that has been proposed is that $x(t)$ is a vector of real-valued numbers but with weights to be fuzzy sets $\left(\overline{W}_{i,j,k}\right)$. This particular form is sometimes useless for practical problems from the viewpoint of fuzzy uncertainty in that it is not necessary to deal with a non-fuzzy problem using the fuzzy logic approach. The exception may be that the problem itself is fuzzy, but is represented by a real-valued vector of membership degrees. Even in this case the rationality of this approach is still open to argument.

Ishibuchi[4] proposed an architecture of fuzzy neural networks for the input fuzzy sets to one output fuzzy set (that is, two-class classification), assuming that these fuzzy sets are all fuzzy numbers represented by real number intervals, but with learning on the real weights (the aberrant form of FNN_3) performed by σ-cuts on the signals ($\sigma = 0.2, 0.4, 0.6, 1.0$). It has been shown in (8.27) that this approach is too complicated and sometimes fails. In fact, if this architecture includes more than one output node (multi-class classification), the cost function of the network defined in (4) will be too complicated to easily derive an explicit algorithm according to the proposed learning mechanism. Hayashi[27] discussed the standard form of FNN_3, assuming that all the signals (input fuzzy sets and output fuzzy sets) are fuzzy subsets of [0, 1], but the weights may be fuzzy sets of real numbers. In order to train the FNN_3, they proposed a learning algorithm, called the "fuzzified delta rule." This rule is only derived by replacing fuzzy differentiation with real differentiation. Both fuzzy equations and fuzzy sets are "forgetful" substituted by the real equations and the real points in the derivation, respectively. This rule is short of theoretic support, and seems to be just a "guess." The exact performance of this rule as applied in fuzzy expert systems, fuzzy hierarchical analysis, and fuzzy systems modeling from (8.28) is not known. In (8.29), the authors propose to adapt the parameter space of the FNN_3 by using the genertic algorithm techniques.

8.5 CONCLUSION

The study reported in this paper is a contribution towards the fundamental issues of fuzzy neural computations. First, a geometrical interpretation of fuzzy neuronal operations was presented using a fuzzy-set-as-a-vector viewpoint. Then, a generalized morphology of computational neurons was proposed to show the common characteristics of information processing and the various motivations between fuzzy neuronal models and other neuro-computing models. Based on the geometrical interpretation and the proposed generalized morphology, it is clear that a fuzzy neuronal processor has the same mechanism of information processing as that of the normal neuronal models, but is constructed by a different implementation (operational logic). With this different kind of operational aspects of neuronal computations, fuzzy neural models can directly deal with practical problems with linguistic vagueness and data imprecision.

After having established some foundations of fuzzy neural computations in this chapter, further work is in progress in this area. This includes studies on the dynamics of fuzzy neuronal models, the architecture of fuzzy neural systems, the cognitive evolution through fuzzy neural networks and genetic algorithms, and their industrial applications.

REFERENCES

1. Kosko, B. *Neural networks and fuzzy systems—a dynamical systems approach to machine intelligence.* Prentice Hall (1992).

2. Gupta, M. M. and J. Qi. "On fuzzy neuron models." *IJCNN'91.* Vol. II, Seattle (1991), pp. 431-434.

3. Kuncicky, D.C. and A. Kandel. "A fuzzy interpretation of neural networks." *Proc. 3rd IFSA Congress.* Seattle (1989), pp. 113-116.

4. Ishibuchi, H., et al. "An architecture of neural networks for input vectors of fuzzy

numbers." *Proc. IEEE Int. Conf. on Fuzzy Systems.* San Diego (1992), pp. 1293-1300.

5. Cohen, M.E. and D. L. Hudson. "Approaches to the handling of fuzzy input data in neural networks." *Proc. IEEE Int. Conf. on Fuzzy Systems.* San Diego (1992), pp. 93-l00.

6. Yamakawa, T. and D.L. Furukawa. "A design algorithm of membership functions for a fuzzy neuron using example-based learning." *Proc. IEEE Int. Conf. on Fuzzy Systems.* San Diego (1992), pp. 75-82.

7. Yang, S Z. and H. Ding. *Knowledge-based diagnostic reasoning.* Qianghai University Press (l993).

8. Pal, S.K. and S. Mitra. *Multilayer perceptron, fuzzy sets and classification.* IEEE, NN-3 (1992), pp. 683-697.

9. Gupta, M.M. and G.K. Knopf. "Fuzzy neural network approach to control systems." *Int. Symposium on Uncertainty Modeling and Analysis.* University of Maryland, IEEE Computer Society Press (1990), pp. 483-488.

10. Gupta, M.M. *Special series of lectures on neural computing systems.* College of Engineering, University of Saskatchewan, Canada, 1992.

11. Gupta, M.M. "Fuzzy logic and neural networks." *Proc. 2nd Int. Conf. on Fuzzy Logic and Neural Networks* (IIZUKA'92), Vol.1 (1992), pp. 157-160.

12. Simpson, P.K. "Fuzzy min-max neural networks - Part l: Classification." *IEEE Trans. Neural Networks.* Vol. 3, No. 5 (1992), pp. 776-786.

13. Keller, J.M., et al. "Neural network implementation of fuzzy logic." *Fuzzy Sets and Systems*, Vol. 45 (1992), pp. l-12.

14. Lippmann, R.P. *Introduction to computing with neural nets.* IEEE ASSP-4 (1987), pp. 4-12.

15. Sanner, R.M and Jean-J.E. Slotine. "Gaussian networks for direct adaptive control." *IEEE Trans. Neural Networks.* Vol. 3 (1992), pp. 837-864.

16. Pedrycz, W. "A referential scheme of fuzzy decision making and its neural networks structure." *IEEE Trans. Sys., Man and Cyner.* Vol. 21 (1991), pp. 1593-1604.

17. Ding H. and M.M. Gupta. "Adaptive aggregation in decision making: a fuzzy neural approach." *SPIE Int. Conf. on Applications of Fuzzy Logic Technology.* Boston, USA (1993).

18. Zimmermann, H.J. *Fuzzy set theory and its applications.* Kluwer Academic Publishers, l985, 1991.

19. Yager, R.R. "Explanatory models of expert systems." *Int. J. of Man-Machine Studies*, Vol. 23 (1985), pp. 539-549.

20. Wang, S.R and Ayeb, B.E., "Diagnosis hypothetical reasoning with a competition-based neural architecture." *IJCNN'92*. Vol. I (1992), pp. 7-12.

21. Yager, R.R. "OWA neurons: a new class of fuzzy neurons." *IJCNN'92*. Vol. I (1992), pp. 226-231.

22. Lin, C.T. and C.S. George Lee. "Neural network-based fuzzy logic control and decision system." *IEEE C-40* (1991), pp. 1321-1336.

23. Okada, H., et al. "Initializing multilayer neural networks with fuzzy logic." *IJCNN'92*. Vol. I (1992), pp. 239-244.

24. Yager, R.R. "On a semantics for neural networks based on fuzzy quantifiers." *Int. J of Intelligent Systems,* Vol. 7 (1992), pp. 765-786.

25. Baldwin, J.F. and N.C. Guild. "Feasible algorithms for approximate reasoning using fuzzy logic." *Fuzzy Sets and Systems*. Vol. 3, No. 3 (1980), pp. 225-251.

26. Homenda, W. and W. Pedrycz. "Processing of uncertain information in a linear space of fuzzy sets." *Fuzzy Sets and Systems*. Vol. 44, No. 2 (1991), pp. 187-198.

27. Hayashi, Y., et al. "Fuzzy neural networks with fuzzy signals and weights." *IJCNN'92*. Vol. II (1992), pp. 696-701.

28. Gupta, M.M. and H. Ding. "Learning in fuzzy-sets neural networks by genetic algorithms" (in preparation).

Appendix to Chapter 8

A8.1 FUZZY SPACE AND FUZZY VECTORS

In fuzzy-set theory, the given fuzzy sets are usually assumed to come from the same universes of discourse. In order to deal with the fuzzy sets from the different universes of discourse the *extension principle* may be used, but this is often difficult to do. An alternative view of fuzzy sets related to the fuzzy neural computations is presented in the following section.

A8.1.1 Fuzzy Vector Space

Let $D = \{D_1 D_2 ... D_n\}$ be n universes of discourse, and suppose that a fuzzy vector space S_j^i ($i = 1, 2, ... n; j = 1, 2, .., m_i$) is defined on D_i by its membership function $\mu_{i,j}$ taking values on the unit interval [0, 1], that is, $\mu_{i,j}: D_i \rightarrow [0, 1]$. Then we give the following definition for a fuzzy vector space S.

Definition 1 S is a fuzzy vector space if it satisfies the following rules:
(a). There exists a binary operation in S, namely addition and denoted by "(+)", for all $x, y, w \in S$ such that

    i) $(x (+) Y) (+) W = x (+) (Y (+) W)$ (associative law)
    ii) $x (+) Y = Y (+) x$ (commutability)
    iii) $x (+) 0 = x$ (zero element)

(b). There exists a scalar multiplication operation in which every fuzzy scalar $a, b \in [0, 1]$ can be combined with every element of the fuzzy vector $x \in S$ to give an element, that is, $a \otimes x \in S$ for all $x, x \in S$ such that

    i) $a \otimes (b \otimes x) = (a \otimes b) \otimes x$ (associative law)
    ii) $(a (+) b) \otimes x = a \otimes x (+) b \otimes x$ (distributive law)
    iii) $a \otimes (x (+) Y) = a \otimes x (+) a \otimes Y$ (associative law)
    iv) $1 \otimes x = x$ (unit element)
    v) $0 \otimes x = 0$ (zero element)

It should be noted that i) compared to the definition of the traditional real or complex number vector space, this kind of fuzzy vector space has no negative elements in it; and ii) S is usually a finite space described below.

A8.1.2 Fuzzy Vectors

Definition 2 A fuzzy vector $\mathbf{x}$ is an ordered set of fuzzy element, denoted as $x = \{\mu_{i,j}(x_k^i)\}$, where $\mu_{i,j}(x_k^i) \in [0, 1]$ ($i = 1, 2, ..., n; j = 1, 2, ..., m_i; k = 1, 2, ..., K_i$) is the membership

Foundations of Fuzzy Neural Computations

degree of x_k^i in S_j^i, and n is the number of universes of discourse; m_i is the number of fuzzy sets in the i-th universes of discourse and K_i is the number of fuzzy variables which are defined in the i-th universes of discourse.

In practice, the following two forms of fuzzy vectors are encountered:

Type one: $\quad x_I = (\mu_{1,j}(x_k^1), \mu_{2,j}(x_k^2), ..., \mu_{n,j}(x_k^n))^T$

Type two: $\quad x_{II} = (\mu_{i,1}(x_k^i), \mu_{i,2}(x_k^i), ..., \mu_{i,m}(x_k^i))^T$

As an example, suppose that $D = \{D_1, D_2\}$, where $D_1 = \{0, 1, ... 200\}$, the fuzzy sets, *'medium'* for medium and *'heavy'* for large of traffic densities which are evaluated by counting the number of passing cars per minute and take values in the interval $[0, 1]$. This may be represented as

$$S_{medium}^l = \exp(-(x-100)^2/100), \quad x \in D_l$$

and

$$S_{heavy}^l = \exp(-(x-200)^2/5000), \quad x \in D_l$$

Now, depict $D_2 = \{-50, 49, ..., 50\}$. The fuzzy set *'extremely cold'* for the temperature of extremely cold weather may be represented as

$$S = \begin{cases} \exp(-(x+40)^2) & x \in [-40, 50]) \\ 1 & x \in [-50, -40] \end{cases}$$

Now consider:

$$x_I = [\mu_{I,1}(90), \mu_{2,1}(-40)]^T = [0.368, 1]^T$$
$$\mathbf{x}_{II} = [\mu_{I,1}(100), \mu_{1,2}(100)]^T = [1, 0.135]^T$$

one may be asked to determine the time-length of traffic light in each case. Besides, a fuzzy vector which is the combination of the above two forms of vectors may also be given. The processing for this combined vector can easily be derived from those for each form. It should also be noted that:

i) If $x = (x_1, x_2, ..., x_n)^T$, type one or type two, is an element of S, then S is called an n-dimensional fuzzy vector space, where $x_i \in [0, 1]$. Each fuzzy element in x corresponds to a coordinate in S and the range of each coordinate is within the unit interval $[0, 1]$. Therefore, the fuzzy vector space is a solid (or continuous) unit hypercube, denoted by $I^n = [0, 1]^n$;

ii) A fuzzy vector $x \in S$ usually has its start-point at the origin $(0, 0, ..., 0)$ and directs to an end-point $(x_1, x_2, ..., x_n)$ in S, as shown in Figure 8.7.

The notion of such a fuzzy vector space has wide applications in pattern recognition, neural computing and measure theory. For example, in pattern classification problems, it is possible to describe the degree to which a sample input belongs to a feature set. Each pattern class is represented by a fuzzy hypercube space.[12]

A8.2 FUZZY VECTOR OPERATIONS

Let $x = (x_1, x_2, ..., x_n)^T \in S^n$, $y = (y_1, y_2, ..., y_m)^T \in S^m$, then the following vector operations are defined:

A8.2.1 Basic Operations

Suppose that n is equals to m, that is, $S^n = S^m$, to be the same n-dimensional fuzzy space,. Then, the fuzzy *MIN-MAX* operators are employed to define basic operations between fuzzy vectors. These fuzzy operations are basic in the development of neuro-fuzzy mathematics employed in this paper.

(i) **Vector addition**

$$x (+) y = ((+) (x_i, y_i) \mid i = 1, 2, ..., n)^T = (MAX\{x_i, y_i\} \mid i = 1, 2, ..., n)^T \quad (8.36)$$

(ii) **Vector subtraction**

$$x (-) y = ((-) (x_i, y_i) \mid i = 1, 2, ..., n)^T = (MIN\{x_i, y_i\} \mid i = 1, 2, ..., n)^T \quad (8.37)$$

(iii) **Scalar combination**

$$x' = \lambda \otimes x = (\otimes (\lambda, x_i) \mid i = 1, 2, ..., n)^T = (MIN\{1, \lambda x_i\} \mid i = 1, 2, ..., n)^T \quad (8.38)$$

for any scalar $\lambda \in [0, \infty]$

(iv) **Inner product**

$$x(\cdot)y = MAX \{MIN(x_i, y_i) \mid i = 1, 2, ..., n\} \quad (8.39)$$

Figure 8.9 gives an intuitive geometric illustration of the above basic operations. As an example, suppose $x = (0.1, 0.2, 0.3, 0.5)^T$, $y = (0.4, 0.3, 0.5, 0.1)^T$, $\lambda = 5.0$. Then

$x (+) y = (0.4, 0.3, 0.5, 0.5)^T$,
$x (-) y = (0.1, 0.2, 0.3, 0.1)^T$,
$x' = \lambda \otimes x = (0.5, 1, 1, 1)^T$,
$x (\cdot) y = 0.3$.

It should be noted that we know the *MIN-MAX* operation is a kind of most pessimistic-optimistic operations and suffers from several problems.[13] Many other kinds of t-norms and t-conorms have been defined, which could be employed here, such as using the algebraic operation. For example,

Foundations of Fuzzy Neural Computations

$$x (+) y = ((x_i + y_i - x_i y_i) \mid i = 1, 2, ..., n)^T \quad (8.40)$$
$$x (-) y = ((x_i \cdot y_i) \mid i = 1, 2, ..., n)^T \quad (8.41)$$

A8.2.2 Fuzzy Matrix

Definition 3 A n by m fuzzy matrix M is a rectangular array of fuzzy elements m_{ij} arranged in n rows and m columns, denoted by $M = [m_{ij}]_{n \times m}$, where $m_{ij} \in [0, 1]$.
Further more, it can be defined:

(v) Vector matrix multiplication

$$y = M \text{ (o) } x = (MAX\{MIN\{m_{ij}, x_i\} \mid i=1, 2, .., n; j=1, 2, .., m\})^T \quad (8.42)$$

(vi) Correlation product

$$M = x \copyright y^T, \, m_{ij} = MIN\{x_i, y_i\}, \, i = 1, 2, ..., n, j = 1, 2, .., m \quad (8.43)$$

It should be noted that,

i) a row matrix x_r is an $1 \times m$ fuzzy matrix, e.g., $x_r = [x_{11}, x_{12}, ..., x_{1m}]$, while a column matrix x_c is an $n \times 1$ fuzzy matrix, e.g., $x_c = [x_{11}, x_{21}, ..., x_{n1}]^T$ and a square fuzzy matrix is a $n \times n$ fuzzy matrix;

ii) a column matrix x_c can be considered to be an n-dimensional fuzzy vector in I^n, while $n \times m$ matrix M to be an extended $n \times m$-dimensional fuzzy vector in $I^{n \times m}$.

9

A CONTROL ALGORITHM FOR KNOWLEDGE-BASED TEXTURE IMAGE SEGMENTATION

Z. Zhang, Z. Lang, and R.E. Scarberry
Medical University of South Carolina, USA

M. Simaan, University of Pittsburgh, USA

Most numerical image segmentation and interpretation techniques rely heavily on local image properties to make classification decisions. In a knowledge-based image segmentation system, however, the classification of one image region is often related to the classification of other regions in the image. It is necessary for such systems to have an effective control mechanism that (1) coordinates and balances the segmentation process over the entire image so that the final results are less dependent on the order in which the image is processed; and (2) provides the means to deliver information and knowledge of diverse sources to the decision making process at different spatial resolution levels. In this chapter, we present such a control mechanism for knowledge-based 2D/3D image segmentation based on an iterative special data structure construction algorithm. We prove that as long as the stopping criteria used at each of the spatial data structure expansion iterations are consistent and in an ordered sequence, the control algorithm will work for any arbitrary images. Results from prototype systems that employed this control mechanism demonstrate d the effectiveness of the algorithm using 2D and 3D real image data.

9.1 INTRODUCTION

Image segmentation is the process "that subdivides an image into its constituent parts or objects."[1] This is achieved either through aggregation of neighboring image areas of similar properties, or by decomposition of large heterogeneous image areas. In many image interpretation applications, such as biomedical imaging and geological exploration, the segmentation is typically a supervised process where all the possible types of image regions are known in advance. In this case, image segmentation becomes a spatially-oriented classification and labeling process. Over the past two decades or so, numerous techniques for supervised image segmentation have been suggested in the literature, all with the essential attempt to integrate the inherent spatial relations/constraints in an image into the classification decision making process of individual areas. For example, the so-called region growing methods take advantage of the fact that neighboring areas in a natural image are more likely to belong to the same image region.[2] However, these relations/constraints are mostly domain-independent knowledge of images in general. The application of domain-specific knowledge in an early vision processing such as image segmentation is, however, often a much more difficult task. In most reported image interpretation systems, expert knowledge is invoked only after an image has been initially partitioned into segments, and is used to help in labeling and grouping the segments to arrive at a final interpretation.

The major obstacles here, as indicated by Zucker,[3] may be due to the practical difficulties involved in devising an adequate control mechanism that is capable of *delivering the right piece of knowledge to the right place at the right time.* To be more specific, two important issues arise when knowledge is applied during, rather than after, the segmentation process. First, the introduction of human knowledge, especially of the type that enforces semantic constraints on adjacent image blocks, might result in a final segmentation that is dependent on the order in which the image is processed. Second, the knowledge and information used to perform segmentation are typically from diverse sources and in various forms (numerical and/or symbolic); and may refer to situations of different spatial scales. The unification and actual utilization of such information in classification decision making can be extremely challenging.

Previous efforts in this area have been very limited in the image processing literature. In the mid-seventies, Feldman and Yakimovsky,[4] and Tenenbaum and Borrow[5] separately reported on two similar systems which arrive at a final segmentation of an image by repeatedly merging neighboring regions starting with an initial partition. Heuristic knowledge is used either to help calculate "boundary strengths" between pairs of adjacent regions so that regions sharing the weakest boundary merge first, or to maintain a list of possible interpretations for pairs of neighboring image regions so that the ones sharing the most common interpretations can merge first. In 1984, Nazif and Levine[6] reported on a knowledge-based system devoted to low-level image segmentation. The system uses multi-layers of controlling rules concerning strategy selection and execution, and domain-independent image segmentation. In 1987, Zhang and Simaan[7] reported on a rule-based interpretation system for segmentation of seismic images guided by a centrally controlled region growing scheme.

In this chapter, we present a control mechanism for knowledge-based segmentation of two or three dimensional (2D/3D) image data. This control mechanism, based on an **I**terative **S**patial **D**ata structure **E**xpansion and **L**ink (ISDEL) algorithm, (i) coordinates all the spatial image data classification processes during the segmentation such that the possible impact of processing order-dependency is minimized; and (ii) provides the necessary means for the integration of domain-specific knowledge and numeric information from image data during the segmentation.

In what follows, Section 9.2 gives a formal presentation of a generic form of the ISDEL control algorithm, and discusses its advantages in terms of the two issues we just mentioned above. Section 9.3 provides details of how symbolic knowledge and numeric information from image data are integrated during the parallel knowledge-based image data classification process and how delayed decision making helps to reduce the impact of processing-order dependency problem. In Section 9.4, two actual prototype implementations of our algorithm are presented for applications of (1) outdoor scene segmentation; and (2) 3D biomedical image data segmentation. The material in this paper extends the discussion in a previous paper.[8]

9.2 AN ITERATIVE SPATIAL DATA STRUCTURE CONSTRUCTION ALGORITHM FOR KNOWLEDGE-BASED IMAGE DATA SEGMENTATION

9.2.1 Spatial Data Structure: Quadtree and Octree

In order to perform spatial-oriented classification (supervised segmentation) of image data, and to apply domain-dependent knowledge regarding spatial relations and constraints among image regions, one of the fundamental problem is to represent the image data in a spatial data structure that allows easy access to and manipulation of the image data at different spatial resolution levels.

Quadtree and Octree have long been used in 2D/3D image data compression/storage and processing.[9] These data structures are characterized by recursively decomposing a multi-dimensional data blocks into a fixed number of equal size sub-blocks (4 quadrants for 2D data or 8 octants for 3D data, etc.). The resultant sub-blocks are organized as nodes into a tree data structure with the original image data block as its root. During the decomposition, a stopping criterion decides whether or not a particular node (block) needs to be further divided. The recursive decomposition stops only when the image data in every leaf node block have satisfied the stopping criterion. For binary image data, this criterion is often a test to see whether the image block consists of pure 0's or 1's. In the case of gray level images, the criterion could be a threshold on some kind of uniformity measures of the image data within the block. As one may easily see, the final representation of a given image is largely determined by the stopping criterion and a more "stringent criterion" will result in a more fully expanded quadtree/octree representation of the image.

Our control algorithm for knowledge-based image data segmentation is based on a process in which a spatial data structure (a tree) representation of the image data is constructed in an iterative fashion using a sequence of stopping criteria that become more and more stringent. Knowledge-based image data classifications are performed during the iterations at different spatial resolution levels. In order to present the control algorithm in a more general form, we need to have a more formal treatment of the relationship between stopping criteria and the resultant tree representations of the image data. For the sake of simplicity, in the following discussion, the generic term *tree* is used which may be replaced by *quadtree* for 2D data and *octree* for 3D data.

Definition 1 *A stopping criterion $R(\cdot)$ is a Boolean function. For a given leaf node n in a tree, $R(n) = true$ means that the node should not be further decomposed under the criterion R.*

Definition 2 *Let $T(I, R)$ denote the tree representation of a given image I constructed using a stopping criterion R, and $N_{T(I,R)}$ be the collection of all nodes in $T(I, R)$. A stopping criterion R_1 is said to be more stringent than another stopping criterion R_2 with respect to a given image I, if and only if $N_{T(I,R_2)} \subset N_{T(I,R_1)}$. We write such a relationship as $R_1 \succ R_2$, or equivalently, $R_2 \prec R_1$, w.r.t. I. Two stopping criteria are said to be equally stringent, written as $R_1 = R_2$, when $N_{T(I,R_1)} = N_{T(I,R_2)}$.*

A subset of $N_{T(I,R)}$ would be the collection of all leaf nodes in $T(I, R)$ and is denoted by $L_{T(I,R)}$. Note that for any node $n \in L_{T(I,R)}$, $R(n) = true$, and for any node $n \in N_{T(I,R)}$ and $n \notin L_{T(I,R)}$, $R(n) = false$.

Corollary 1 *The relationship of "more stringent" satisfies the law of transitivity, i.e. if $R_1 \succ R_2$ and $R_2 \succ R_3$ w.r.t. I, we then have $R_1 \succ R_3$ w.r.t. I.*

Proof. $N_{T(I,R_2)} \subset N_{T(I,R_1)}$ and $N_{T(I,R_3)} \subset N_{T(I,R_2)} \Rightarrow N_{T(I,R_3)} \subset N_{T(I,R_1)}$.

Definition 3 *A class of stopping criteria $\{R_i\}$ are said to be consistent if and only if for any arbitrary image I and any two criteria R_i and R_j from $\{R_i\}$, $R_i \neq R_j$, only one of the following is true: $R_i \succ R_j$, $R_i \prec R_j$, or $R_i = R_j$ w.r.t. I (the law of trichotomy).*

Corollary 2 *If two stopping criterion R_1 and R_2 are from a class of consistent stopping criteria and $R_1 \succ R_2$ w.r.t. an image I, then $R_1 \succ R_2$ with respect to any arbitrary images. In this case, R_1 is said to be consistently more stringent then R_2.*

Proof. If there exist two images I_1 and I_2 such that $R_1 \succ R_2$ w.r.t. I_1 and $R_1 \prec R_2$ w.r.t. I_2, then for the image $I = I_1 \cup I_2$, we will have $R_1 \not\succ R_2$ and $R_1 \not\prec R_2$ w.r.t. I. This contradicts the assumption that both R_1 and R_2 are from a class of consistent stopping criteria.

The above corollary and the properties of transitivity and trichotomy ensure that a class of consistent stopping criteria can be linearly ordered according to their relative "stringence."

Let $V(n)$ denote the image block represented by node n in a tree. Then for any tree data structure representation of any image I, we always have $I = V(root)$.

Corollary 3 *For any tree representation T (I, R) of any arbitrary image I constructed using a certain stopping criterion R, we have*

$$\bigcup_{\forall n, n \in L_{T(I,R)}} V(n) = I \quad (9.1)$$

and for any two leaf nodes $n, m \in L_{T(I,R)}, n \neq m$

$$V(n) \cap V(m) = \Phi . \quad (9.2)$$

Proof. Clear from the recursive construction process of the tree representation of the image as discussed in the beginning of this section. This corollary states that the collection of image blocks represented by the set of all leaf nodes forms a non-overlapping partition of the original image data.

9.2.2 ISDEL Algorithm for Controlling Knowledge-based Image Data Segmentation

From the above discussion, we can see that if we use an ordered sequence of consistent stopping criteria $\{R_i : R_i \prec R_{i+1}, i = 0, 1, \ldots N\}$ to iteratively expand a tree representation of an input image I, we will be able to obtain a sequence of leaf node sets $\{L_{T(I,R_i)}\}$, each corresponding to a non-overlapping partition of I (Corollary 3) and with increasingly finer spatial resolution levels (Corollary 1). Corollary 2 ensures that such an iterative expansion procedure, once established, may be applied to any arbitrary input images of a given type.

To formulate such an iterative algorithm for controlling knowledge-based image data segmentation, we need first to define a class of consistent and increasingly more stringent stopping criteria. As in many other image segmentation methods, we take advantage of the fact that for most natural images, neighboring areas of uniform image properties are more likely to belong to the same image region and should be classified as a whole. Let $P(n)$ be a measure of uniformity computed over the image data represented by the node n in a tree data structure (In cases where image element values can not be used directly for segmentation purpose, such as in texture images, $P(n)$ will have to be computed using some secondary image attributes). The class of stopping criteria $\{R_{P,Q_i}(\cdot)\}$ then may be defined as:

$$R_{P,Q_i}(\cdot) \begin{cases} \text{true} & f\ P(n) \geq Q_i, \text{ or} \\ & \text{size } n \leq \text{MinResolution, or} \\ & n \text{ has been permanently marked} \\ & \text{off from further expansion,} \\ \text{false} & \text{Otherwise} \end{cases}, \quad (9.3)$$

where $\{Q_i\}$ is a sequence of monotonically increasing thresholds on uniformity measures.

It is easy to verify that $\{R_{P,Q_i}(\cdot)\}$ forms an ordered class of stopping criteria for tree representation of any given image data as long as the uniformity measure $P(\cdot)$ is *single valued*.

Algorithm ISDEL:

/* an **I**terative **S**patial **D**ata structure **E**xpansion and **L**inking algorithm for controling knowledge-based image segmentation */

User supplied parameters:

    a. Initial uniformity measure threshold Q_{init};
    b. Threshold increment $\Delta Q_i = f(i, segmentation\_status)$;
    c. Maximum value for uniformity threshold Q_{max};
    d. Minimum spatial resolution for node expansion (used in $R_{P,Q_i}(\cdot)$).

1. initialization:

    a. $Q_0 = Q_{init}$;
    b. Create a tree T_0 with only one node, the root r, such that $V(r) = I$ (image to be segmented);
    c. *Forced_Classification_Flag* = False.

2. Recursively expand the tree until $T_i = T(I, R_{P,Q_i})$ (i.e. all leaf nodes on the tree have satisfied the stopping criteria).

3. $Q_{i+1} = Q_i + \Delta Q_i$. If $Q_{i+1} > Q_{max}$, *Forced_Classification_Flag* = True.

4. Perform **Knowledge-based Leaf Node Classification** in parallel on each leaf node $n \in L_{T_i}$ that has not been *Positively Classified*.

5. Link all neighboring leaf nodes that have been *Positively Classified* to the same image regions;

6. Update the **Global Segmentation Database**; and compute ΔQ_{i+1}.

7. If all leaf nodes have been *Positively Classified* and permanently marked off from further expansion, output result and exit.

8. Go to step **2**.

In step 4 of the above algorithm, knowledge-based leaf node classification, information and knowledge regarding the membership of the image data represented by the leaf node to possible image regions are pulled together from diverse sources. When the *Forced_Classification_Flag* is set false, the classification decision is usually a tentative one unless the accumulated evidence has become overwhelmingly compelling. In that case, the node will be marked as *Positive Classified* and will not enter any further expansions. However, if the *Forced_Classification_Flag* is set true, positive classifications are made on every leaf node based on the best information and knowledge available. The **Global Segmentation Database** contains uptodate descriptions of the current segmentation status. It provides the necessary information for applying certain domain-dependent global constraints among image regions. In order to minimize the processing order-dependence problem, the database is only updated once in each iteration, in step 6. after the leaf node classification and linking.

9.2.3 Advantages of the ISDEL Algorithm

In the ISDEL algorithm, the actual application of knowledge and the classification of image data take place only in steps 4-6. The procedures involved in these steps will be discussed in the next section where the integration of knowledge and information from diverse sources is explained in detail. However, The process of stepwise expanding a hierarchical spatial data structure (*a tree*) representation of the image data while cumulatively collecting information and performing knowledge-based classification renders its own merits in terms of fulfilling exactly the two requirements needed for a control mechanism for knowledge-based image segmentation: (i) helping the delivery and application of knowledge; and (ii) coordinating image data classification processes to reduce processing-order dependence of the final segmentation result.

First, ISDEL provides a controlled environment for accessing and manipulating image data at different spatial scales with approximately the same level of image property uniformity. Application of domain-specific knowledge, such as constraints on probable/improbable concurrence of image events of various spatial scales, becomes much easier. For example, rules concerning global conditions may be introduced during the early stages of the segmentation and have influence over classification decisions of image components at finer spatial scales later. Computationally, many existing fast tree traverse-based algorithms[9] make tasks such as verifying classification status of neighboring image areas and grouping similar image areas into large regions much more tractable.

Secondly, one of the advantages of manipulating image data using a hierarchical data structure is that local operations can have effects of a global scale. For instance, in many practical problems, image data often consist of large regions with highly uniform image properties. These regions, under the ISDEL algorithm, will be classified in the early stage of segmentation in an extremely efficient "whole sale" manner.

Under ISDEL, knowledge-based classifications of leaf nodes are performed *in parallel.* Although rules enforcing semantic constraints on adjacent image areas are used in classification decision making, the information which triggers the firing of such rules is all based upon evidence accumulated prior to the current iteration. All nodes are independently processed and the results are compiled only once at the end of each iteration into the **global segmentation database.** All these measures are aimed at minimizing the impact of processing-order dependence.

Marr[10] had suggested two principles for designing visual information systems. One of them is the *Principle of Least Commitment* which requires that a system should not do something that may later have to be undone. In ISDEL, this is realized by having mostly tentative classification decisions during the early stages of the segmentation. As the iteration continues, more information is gathered for leaf nodes of better data uniformity and finer spatial resolution. Definite classifications then become possible. The adherence to the *Principle of Least Commitment* serves again the purpose of resolving the problem of processing-order dependence since the less irrevocable a bad decisions is made, the better chance it will to be corrected by the gathering of additional information.

Lastly, the idea of segmenting an image by classifying image components of strong evidence first and then using the results to guide further analysis of fine details is intuitively, and in a very simplified sense, close to the human visual perception process. The region growing method, which is based on such an idea, has been shown to be very robust as long as the starting seeds are correctly classified. The ISDEL algorithm may be considered as a "structured" version of region growing where classifications are performed on data organized in a hierarchical tree structure and the "Positively Classified" nodes in early stage of the expansion helps to guide through the classification of leaf nodes of finer spatial resolutions later.

9.2.4 Uniformity Measures

There have been a number of uniformity measures suggested in the literature.[10] Most of these measures are based on some type of statistics of the image data or secondary image attributes computed from the image data. It is expected that when the size of the image data represented by the leaf nodes becomes small, the uniformity measures will become unstable and less meaningful. However, as long as the uniformity measure chosen are single-valued for any given node with its corresponding image data, the ISDEL algorithm should still work fine.

ISDEL does not mandate the use of any particular uniformity measure. The selection of such a measure should depend largely on the particular type of image data being processed. For the prototype systems we have developed (see examples in Section 9.4), the image data are of texture nature. For such data, we defined a *Coarseness Measure* to measure the relative improvement of uniformity of texture feature measures if an image block is split into sub-blocks and being considered separately. In the following definition, the general term *image* refers to either the image data or computed secondary image attributes (feature measures), and could be either 2D or 3D.

Let A denote a block in an image I and let p be a partition of A that divides the block into N sub-blocks $A_1, A_2, \ldots, A_N$ such that $A = \cup_{i=1}^{N} A_i$. Let σ and m be the standard deviation and mean value of the image data within block A respectively, and σ_i and m_i be the standard deviation and mean value respectively, of image data in $A_i, i = 1, 2, \ldots, N$. Then the coarseness measure CM of image I computed within block A with respect to partition P is defined as:

$$CM = \sigma^2 - \frac{1}{|A|} \sum_{i=1}^{N} |A_i| \sigma_i^2. \tag{9.4}$$

With some simple derivations,

$$\sigma^2 = \frac{1}{|A|} \sum_{i=1}^{N} |A_i| [\sigma_i^2 + (m_i - m)^2]. \qquad (9.5)$$

Substituting (5) into (4) yields,

$$CM = \frac{1}{|A|} \sum_{i=1}^{N} |A_i| \cdot (m_i - m)^2$$

$$= \frac{1}{|A|} \sum_{i=1}^{N} |A_i| \cdot (m_i - \sum_{j=1}^{N} \frac{|A_j|}{|A|} \cdot m_j)^2 \qquad (9.6)$$

$$= \frac{1}{|A|} \sum_{i=1}^{N} |A_i| \cdot \left[\sum_{j}^{N} \left(\frac{m_i}{N} - \frac{|A_j|}{|A|} \cdot m_j \right) \right]^2$$

For 2D/3D image data using quadtree/octree for ISDEL:

$$|A_1| = |A_2| = \ldots |A_N| = \frac{|A|}{N}.$$

$$CM = \frac{1}{N^3} \sum_{i=1}^{N} \sum_{j=1}^{N} (m_i - m_j)^2$$

$$= \frac{4}{N^3} \sum_{i=1}^{N} \sum_{j=i+1}^{N} (m_i - m_j)^2 \qquad (9.7)$$

A computational advantage of this *CM* expression is that it concerns only the mean values of image data instead of the standard deviations which are in general much more expensive to compute. With little modification, the *CM* measure can be used in knowledge-based image segmentation under the ISDEL control algorithm since the coarseness measure is simply the opposite of the uniformity measure. For images with multiple computed attributes, a pulled *CM* value can be obtained by taking a weighted sum of *CM* values of individual image attributes.

9.3 INTEGRATION OF KNOWLEDGE AND NUMERIC INFORMATION DURING SEGMENTATION

ISDEL control algorithm provides the necessary mechanism for applying information and knowledge at various spatial scales. However, the actual integration of such information and knowledge, which are often from diverse source and in different forms, is not a simple matter. A lesson learned from the early attempts in building model-based image interpretation systems is that for low-level processing, domain-specific knowledge should not be used in very rigid and exclusive constraint type form. This will either lead to a very specialized system working with only a few idealized problems, or result in a system with an unacceptably complicated control mechanism. Such knowledge should be used in a unified form along with the information contained in the numerical values of the image data. The unification of information of multiple sources should also be achieved in such way that a piece of information or knowledge can only have a measurable amount of influence towards the final result rather than trying to make "yes-or-no" decisions alone. These requirements coincide with Marr's another principle for visual system,[11] *Principle of graceful degradation,* which is to ensure that, whenever possible, degrading the data will not prevent the delivery of at least some of the answer.

In this chapter, we present method based on a *Certainty Factor Vector Updating* procedure. This procedure assumes that it is possible to extract enough information from the image data to form the initial certainty factors regarding the image class membership of the data under consideration. This is often possible in a neighborhood-based image analysis such as texture feature extraction and classification. The following is the general algorithm for knowledge-based classification of a leaf node.

Algorithm KBLNC:
/* **K**nowledge-**B**ased **L**eaf **N**ode **C**lassification procedure to be performed *in parallel* on each unclassified leaf node. */

User Supplied Parameters:
T_D: Threshold for positive classification.

1. Collect local and global information relevant to the classification of the node and place it in a data structure called *INFO*, which includes:

    a. initial certainty factor vector $CF = [cf_1, cf_2, ..., cf_L]$ computed using numerical information contained in the corresponding image data;

    b. information about nearby large image regions that have been positively classified;

    c. information about neighboring nodes on its four sides, if any;

    d. tentative classification and information from parent node, if available;

    e. and other related information regarding the overall segmentation status.

2. Invoke procedure *CF_UPDATE* to perform knowledge-based certainty factor vector updating using information in *INFO*.

3. Calculate a distance measure D from the newly updated certainty factor vector. Classify the node according to the following criterion:

    if $D < T_D$, or *Forced_Classification_Flag* = True, classify the node to the image class which corresponds to the highest *cf* value in the certainty vector, mark the node as *Positively Classified;*

else save the tentative result and relevant information into the node.

In the above algorithm, the certainty factor vector is a vector of L entries, where L is the total number of known possible image classes. An entry cf_i is a number valued in the closed interval [0,1] representing the likelihood of the node under consideration belonging to image class i. A value equal to 1.0 means absolutely positive and 0.0 absolutely negative. The value 0.5 means unknown. The functions that compute the initial certainty factors are very similar to the decision functions in a conventional classification approach except that the functions compute a likelihood number between 0 and 1 instead of offering simple yes-or-no answers.

The procedure *CF_UPDATE* is the most important step in the knowledge-based classification algorithm. In this procedure all information and knowledge accumulated so far are actually used to help making classification decisions. In our prototype systems for 2D/3D texture image segmentation, we used a knowledge-based deduction procedure adopted from Winston's book.[12] In our implementation, the information gathered in the data structure *INFO* is all in the form of individual facts. Each fact has its own certainty factor regarding the relative truthfulness of the fact. The input certainty factor to a rule is the minimum of the certainty factors of all the facts that are the conditions for firing the rule. The output certainty factor of a rule is then the input mapping through a single valued function. For each of the rules in the knowledge base, the mapping function is defined by four parameters cf_0, m, cf_m, and cf_1 in the following way:

$$cf_{output} = \begin{cases} \dfrac{cf_m - cf_0}{m} cf_{input} + cf_0 & \text{if } cf_{input} \leq m \\ \dfrac{cf_1 - cf_m}{1-m}(cf_{input} - m) + cf_m & \text{otherwise} \end{cases} \quad (9.8)$$

It is easy to see that by adjusting the parameters, one can make fine tunings on the behavior and relative power of individual rules.

The formula for combining output from different rules concerning the same conclusions is done through an intermediate variable called certainty ratio r[12] defined as

$$r = \frac{cf}{1 - cf} \quad (9.9)$$

and the inverse transformation is given by

$$cf = \frac{r}{1 + r} \quad (9.10)$$

Assuming that rule 1 and rule 2 support the same conclusion with different output certainty factor values, let r_1 and r_2 be the corresponding certainty ratios, respectively. The combined certainty ratio of the conclusion will then be

$$r_{comb} = r_0 \cdot r_1 \cdot r_2 \tag{9.11}$$

where r_0 is the *a priori* certainty ratio of the conclusion before the update. If the conclusion does not exist before hand, r_0 is normally assumed to be equal to 1.0 (corresponding to an *a priori* certainty factor of 0.5). The final certainty factor of the conclusion is obtained by applying the inverse transformation formula on r_{comb}. Further implementation details and discussions about evidence combination schemes and their effects on an image interpretation system may be found in Zhang and Simaan.[13]

The decision on whether or not a node has accumulated enough evidence and therefore should be *Positive Classified* is based upon comparing a distance measure D computed using the newly updated certainty factor vector against a user specified threshold T_D. For a certainty factor vector of L entries, let

$$cf = max(cf_1, cf_2, \ldots, cf_L), \tag{9.12}$$

$$D_{kmax} = \begin{cases} D_{MAX} - cf_k & \text{if } cf_k < D_{MAX} \\ 0 & \text{otherwise} \end{cases}, \tag{9.13}$$

$$D_{imin} = \begin{cases} cf_i - D_{MIN} & \text{if } cf_k > D_{MIN} \\ 0 & \text{otherwise} \end{cases}, \tag{9.14}$$

where D_{MAX} and D_{MIN} are two user specified upper and lower threshold. The distance D is then defined as:

Definition 4

$$D = \sqrt{D_{kmax}^2 + \sum_{i=1, i \neq k}^{L} D_{imin}^2}. \tag{9.15}$$

This measure is adopted from Love and Simaan[14] and is basically a measure on how far the highest *cf* value in a certainty factor vector has set itself away from the remaining *cf* values. As one might expect that a larger D_{MAX} value and a smaller D_{MIN} will yield a more stringent requirement for making classification decisions.

9.4 EXAMPLE APPLICATIONS

In this section, we provide two examples from our prototype implementations of the ISDEL control algorithm for knowledge-based image data segmentation. One is for the

A Control Algorithm for Knowledge-based Texture Image Segmentation

Figure 9.1. An out-door scene of San Francisco Bay. Four different types of image regions are labeled.

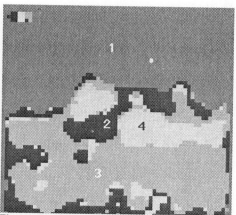

Figure 9.2. Knowledge-based segmentation result using the ISDEL control algorithm. Image regions are labeled the same way as in Figure 9.1.

interpretation of outdoor scene and the other for 3D biomedical image data segmentation. In both systems, subsets of Laws' *Texture Energy Measures* (with extension to 3D data)[15] are used as the secondary numerical image attributes.

The outdoor scene test image of the San Francisco Bay is shown in Figure 9.1. It may be roughly divided into four types of image regions: the sky with clouds (region 1), trees and bushes (region 2), the sea (region 3) and buildings (region 4). Three 2D texture energy features have been selected according to their inter-class discriminatory power: $E5E5$, $E5S5$, and $E5W5$ (all with the macro-mask $ABSAVG$, average of absolute values).[15] The result is given in Figure 9.2 where the corresponding image regions are labeled with the same numbers as in the original image. Since our main emphasis is in the development of the control algorithm, the rules used in processing this image are the simple constraints such as "trees are more likely next to buildings," etc. Considering small size of buildings in the image and the relatively similar appearance of regions, the segmentation offers a reasonable result.

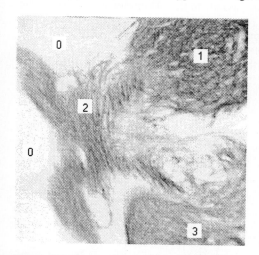

Figure 9.3. Frame #16 of a 3D (256x256x40) confocal fluorescence microscopic image of the right atrio-ventricular (AV) valve portion of a developing chick embryonic heart. Image regions with different muscle texture orientations are numbered 1 to 3. Areas marked 0 are the empty space.

The prototype 3D segmentation system is developed for identifying muscle fiber structures in 3D confocal fluorescence microscopic images. The test data are imaged from the right atrio-ventricular (AV) valve portion of a developing chick embryonic

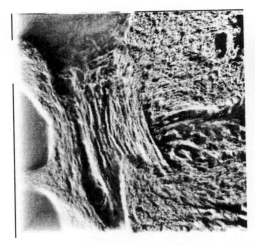
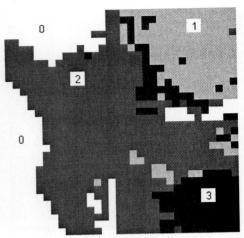

Figure 9.4. Front view of the 3D image data, computed using a Simulated Fluorescence Process (SFP) algorithm.

Figure 9.5. Frame #16 (same as in Figure 9.3) of the knowledge-based 3D segmentation result using an octree version of the ISDEL control algorithm.

heart using a BioRad MRC600 laser scanning confocal microscope. The data consist of forty consecutive 256x256 image frames (extracted from an original data set of size 40x780x512). Figure 9.3 shows one of such 2D frames (No. 16 in the set). In the image, the background marked with number "0", corresponds to the 3D empty space around the muscle fiber tissue. Three different muscle fiber texture patterns are also marked with numbers. In order to provided a better perception of the depth information in the stack of image frames, Figure 9.4 shows the front view of the stack of images, computed using a Simulated Fluorescence Process (SFP) algorithm. Heuristic knowledge used in the system is mainly the ones concerning the continuity and orientation of muscle fibers. Figure 9.5 is the same frame as in Figure 9.3 in the 3D segmentation result obtained using our prototype system. Figure 9.6 is again the same frame, No. 16 of the original images, superimposed with the segmentation boundaries. By mere visual inspection, these boundaries match reasonably well with the original image data except at the very lower portion of the image, which we think might be caused by the effect of the relatively large size of convolution masks used in feature extraction.

9.5 CONCLUSIONS

In this chapter, we have discussed the necessity of having a control mechanism for

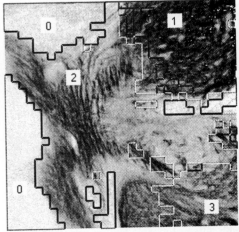

Figure 9.6. Frame #16 superimposed with segmentation boundaries.

knowledge-based image segmentation and presented such a control mechanism based on an iterative spatial data structure construction algorithm ISDEL. We proved that the ISDEL algorithm, once established with a few user specified parameters, will work with any arbitrary image of a given type. Mechanisms which takes advantage of the ISDEL control algorithm for actually incorporating knowledge and information of diverse sources and various spatial scales have been explained in detail. Two examples from prototype systems which we have developed for outdoor scene interpretation and 3D biomedical image data segmentation have demonstrated the effectiveness of our proposed control mechanism.

REFERENCES

1. Gonzalez, R.C., and P. Wintz. *Digital Image Processing.* 2nd ed., Addison-Wesley, Reading, Mass., USA (1987), Chap. 7.

2. Zucker, S.W. "Region growing: childhood and adolescence." *Comp. Graphics Image Processing.* Vol. 5 (1976), pp. 382-399.

3. Zucker, S.W. *Early Vision.* TR-86-5R, McGill University, Canada (1986).

4. Feldman, J.A., and Y. Yakimovsky. "Decision theory and artificial intelligence: I. A semantic-based region analyzer." *Artificial Intelligence.* Vol. 5, (1974), pp. 349-371.

5. Tenenbaum, J.M., and H.G. Barrow. "Experiments in interpretation guided segmentation." *Artificial Intelligence.* Vol. 8, (1977), pp. 241-274.

6. Nazif, A.M., and M.D. Levine, M.D. "Low level image segmentation: an expert system." *IEEE Trans. Pattern Anal. Machine Intell.* Vol. PAMI-6 (1984), pp. 555-577.

7. Zhang, Z., and M. Simaan. "A rule-based interpretation system for segmentation of seismic images." *Pattern Recognition.* Vol. 20, 1987, pp. 45-53.

8. Zhang, Z., M. Simaan, Z. Lang, and R.E. Scarberry. "Controlling Knowledge-based image segmentation using an iterative spatial data structure construction algorithm," to appear in *Computers & Electrical Engineering.*

9. Samet, H. "The Quadtree and related hierarchical data structure." *Computing Surveys.* Vol. 16, No. 2 (1984), pp. 187-260.

10. Levine, M.D., and A.M. Nazif. "Dynamic Measurement of Computer Generated Image Segmentations." *IEEE Trans. Pattern Anal. Machine Intell.* Vol. PAMI-7 (1985), pp. 155-164.

11. Marr, D. "Early processing of visual information." *Philosophical Trans. of Royal Society of London.* B275 (1976), pp. 483-524.

12. Winston, P.H. *Artificial Intelligence.* Addison Wesley, Reading, Mass., USA (1984).

13. Zhang, Z., and M. Simaan. "Knowledge-based reasoning in SEISIS - a rule-based seismic section interpretation system based on texture." Chapt. 6 in *Expert Systems in Exploration.* F. Aminzadeh and M. Simaan eds., SEG Press, Tulsa, OK, USA (1991), pp. 141-159.

14. Love, P.L., and M. Simaan. "Segmentation of a seismic section using image processing and artificial intelligence techniques." *Pattern Recognition.* Vol. 18 (1985), pp. 409-419.

15. Laws, K.I. "Texture image segmentation." *Ph.D. Dissertation.* USC, Ca., USA (1980).

10

KNOWLEDGE-BASED ON-LINE SCHEDULING FOR FLEXIBLE MANUFACTURING

I. Hatono and H. Tamura
Osaka University
Osaka, JAPAN

In this article, first, we describe a brief survey of scheduling problems for production systems under uncertainty. Next, we describe a rule-based on-line scheduling system and an FMS simulation system under uncertainty using stochastic Petri nets, which is developed for creating, debugging, and evaluating the rule base for on-line scheduling. The performance of the rule-based scheduling system is evaluated by comparing with the case in which several fixed priority rules are used.

10.1 INTRODUCTION

Production scheduling[1] has become an important problem for obtaining efficiency and high productivity in highly automated manufacturing systems, such as Flexible Manufacturing Systems (FMS), which consist of numerically controlled machining centers, industrial robots, automated guided vehicles (AGV), material handling systems, and so forth. FMS is much more complex than the other production systems, because:

(1) There exist alternative machines.

(2) It is necessary to generate job transfer schedules, the schedules for operators of machine tools, and the schedules for tool supply, simultaneously.

(3) There exist buffers at each machine tool and some temporal buffers.

In scheduling problems for such complex and large manufacturing systems it is very difficult to find an optimal solution because of computational complexity.

We can classify production scheduling problems into two types: off-line scheduling[2] and on-line (dynamic) scheduling.[3] Off-line scheduling means to decide operation start time of each job (product) for relatively long time span, such as one week, one month, and so on. On the other hand, in on-line scheduling, a job to be processed next must be decided right after the processing of previous job is finished. Therefore, schedulings for long time span are not generated in on-line scheduling.

In flexible manufacturing, in general, products are processed according to a job scheduling obtained with off-line basis. However, in flexible Manufacturing, we often come across the failure of machine tools, a sudden change of production plans, and so forth. In this case, it is necessary to obtain a new schedule immediately. Therefore, off-line scheduling is suitable for "stable manufacturing system," which means a manufacturing system that the failure of machine tools and sudden change of production plans are seldom occurred, and on-line scheduling is suitable for "unstable manufacturing system."

In this chapter, first, we describe several methods for on-line scheduling. Next, we describe a knowledge-based on-line scheduling system. In this system, to cope with the difficulty of on-line scheduling, a rule base for on-line scheduling is developed that dynamically selects appropriate dispatching rules at different time and situations in a manufacturing process based on a knowledge engineering approach.[4] Furthermore, to evaluate the performance of on-line rule base for scheduling, an FMS (Flexible Manufacturing System) simulator is developed by continuous-time and discrete-time stochastic Petri nets [5, 6] with hierarchical structure.

10.2 A BRIEF SURVEY OF SCHEDULING UNDER UNCERTAINTY

We often use (1) OR methods such as branch and bound method[2], (2) scheduling simulation,[7] (3) AI (Artificial Intelligence) methods[8, 9] to solve ordinary scheduling problems. However, it is difficult to apply these methods directly to production systems under uncertainty and dynamic environments because it is necessary to generate schedules in very short time. The methodologies suitable for such production system need to provide the following functions:

(1) Rescheduling,

(2) Stochastic optimization,

(3) On-line (dynamic) scheduling.

Rescheduling is a function such that when the production environment is varied, a new schedule is generated depending upon the variations. The study of timing of rescheduling is reported in.[10] However, it is difficult to apply this method to flexible manufacturing systems because of the unrealistic assumptions such as periodicity of productions.

Stochastic optimization is a function to generate a schedule which optimizes the expected value of a schedule evaluation function by using the stochastic models, such as queuing networks, stochastic Petri nets and Markov chains. However, it is difficult to develop an accurate stochastic model of FMS.

Generally, in on-line scheduling, production processes are scheduled according to an appropriate dispatching rule. In this method, when some jobs can be processed in a resource, a priority rule is fixed throughout the whole process and is applied to select one job to be processed. However, the specific priority rule may not be a good choice over all the situations to resolve conflicts among jobs, because the environment of manufacturing may be changed by failure of a machine tool, by urgent jobs to be processed, and so forth. For other approaches to on-line scheduling, Agnetis et at.[11] reported the method of on-line scheduling by solving a relatively small quasi optimum problem representing an on-line scheduling problem whose objects are limited to AGVs (Automated Guided Vehicles) or job routings in real-time. Kumar and Seidman[12] reported the study about dynamic instabilities and stabilization methods in distributed real-time scheduling. However, since the objects of these approaches are limited, it might be difficult to apply these methods to FMSs. Recently, several AI based approaches to on-line scheduling[4, 13, 14, 15, 16, 17, 18] have been reported. We think AI based approaches are the most practical and useful methods for FMS scheduling because FMS are too complex to apply the other theoretical methods. In the following sections, we describe a rule-based on-line scheduling[4] as an example of AI based approach.

10.3 ON-LINE SCHEDULING USING RULE BASE

Scheduling is conducted under various scheduling objectives such as minimizing flow time, completion time, number of tardy jobs and maximum tardiness time, maximizing production rate, and so forth. This section deals with a rule-based on-line scheduling system for FMS that generates appropriate priority rules to select a job to be fired from a set of conflicting jobs, and then shows an example of the rule base under the given objectives.

10.3.1 A System Structure

To generate a schedule in real-time, it is necessary to resolve three kinds of conflicts as follows:

(1) Selections of alternative machine tools,

(2) Job selections for transporting,

(3) Job selections from an input buffer.

These conflicts are resolved by an appropriate priority rule to achieve a scheduling objective.

To generate a priority rule[19] for resolving conflicts, we construct hierarchically structured rules, each of which has the following form:

$$\text{if } P_i \text{ then } Q_i.$$

A condition P_i is given by a predicate which describes the states of the timed Petri net such as throughput, remaining number of processes, processing times, and an occupation ratio in each buffer, where the occupation ratio at time τ is defined by $\mu_r(p_i)=L(p_i)$ for a buffer place p_i, where μ_r denotes the number of tokens contained in place p_i. Q_i is either a priority rule or a group number of rules to be checked next. The following sixteen predicates describe the states of the stochastic Petri net:

P_1 : *true*: Always true.
P_2 : *false:* Always false.
P_3 : *not (A):* True if A is false; otherwise false.
P_4 : *after (n)* : True if the current time is after time *n*; otherwise false.
P_5 : *buffer1(n,a)* : True if there exists a buffer in *n* processes ahead of the current process such that its occupation ratio is greater than or equal to *a%*; otherwise false.
P_6 : *buffer2(n,a)* : True if the occupation ratios of all buffers in *n* processes ahead of the current process are greater than or equal to *a%*; otherwise false.
P_7 : *buffer3(a):* True if there exists a buffer in all processes after the current process such that its occupation ratio is greater than or equal to *a%*; otherwise false.
P_8 : *buffer4(a):* True if the occupation ratios of all buffers in all processes after the current process are greater than or equal to *a%*; otherwise false.
P_9 : *processing-time(n, a):* True if the ratio between maximum and minimum processing times in *n* processes are equal to *a%*; otherwise false.
P_{10} : *progress(n):* True if there exists a part type in the conflicting part types such that its remaining number of processes is less than or equal to *n*; otherwise false.
P_{11} : *attain:* True if there exists a part type in the conflicting part types such that at least one of its workpieces is finished at the current time; otherwise false.
P_{12} : *best-time1(a):* True if there exists a part type in the conflicting part types such that the ratio between the remaining time to the due date and *Imin,* which means an estimated minimum time to complete all workpiece of one part type, is greater than *a*; otherwise false.
P_{13} : *best-time2(a):* True if ratios between the remaining time to the due date and *Imin* of all conflicting part types are greater than *a*; otherwise false.
P_{14} : *conflict(a):* True if a number of the conflicting part types is greater than *a*; otherwise false.
P_{15} : *machine:* True if a machine operates the current process; otherwise false.
P_{16} : *beginning:* True if there exists the part type that has already started to process in the conflicting part types; otherwise false.

Suppose that a major scheduling objective is JIT (Just-in-Time). Moreover, we consider another objective; reducing the number of set-up, so that each machine tends to process the workpieces of the same part type.

Over a hundred of priority rules have been proposed and investigated through simulation techniques.[7,20,21] In this chapter, we will use the following eleven priority rules:

D_1: Select the job with "earliest due date."

D_2 : Select the job with "shortest slack time."
D_3 : Select the job with "longest tardy time."
D_4 : Select the job with "largest number of tardy jobs."
D_5 : Select the job with "smallest sum of early time."
D_6 : Select the job with "smallest early time ratio."
D_7 : Select the job at random.
D_8 : Select no job.
D_9 : Select the job with "largest occupation ratio in buffer."
D_{10} :Select the job with "smallest occupation ratio in buffer in the next operation."
D_{11} :Select the job with "smallest number of remaining processes."

10.3.2 A Rule Base for the Main Purpose of JIT

Several heuristic knowledges have been found to achieve good performance of the scheduling objective through numerical experiments and examining marking evolutions as follows:

(1) No part type is operated until the operational start time that is calculated by the number of conflicting and uncompleted part types, due date, and the remaining processing time.
(2) Once a part type is started to operate, it is continued to operated even if the condition (1) is not satisfied.
(3) If all the conflicting part types are early to the due date, then a part type with the smallest early time should have a priority.
(4) If there exists the part type that is tardy to the due date in the conflicting part types, then a part type with earliest due date should have a priority.
(5) If the part types are conflicted at a transport, then a part type with smallest occupation ratio in buffer should have a priority.

Each predicate used in a rule base has the following functions. Predicate *after* calculates the operational start time of a first workpiece of each part type. *Buffer* prevents blockings. *Progress* makes a completed workpiece to have a priority for unloading. *Best-time1* and *best-time2* determines whether the operation should be started or not. *Conflict* and *machine* are used to classify the type of conflicts, because the strategies to determine whether the operation should be started or not and to resolve conflicts at a machine tool or transport, are dependent on the number of conflicting part types and the location where the conflicts occur. *Beginning* makes the part type on operation to continue to operate. *Filter* prevents each part type from being operated before its operational start time.

We construct a rule base according to the above heuristic rules. To simplify and speed up of inferences, we translate above heuristic rules into the inference tree shown in Figure 10.1. When the conflicts of transitions occur in the transporting level models, an appropriate priority rule for resolving conflicts is selected by using the rule base in Figure 10.1. In the first stage of the algorithm of resolving conflicts, the first predicate in group 1 in Figure 10.1 is checked. If it is true then the first predicate in a group directed by an arc are examined; otherwise check the next predicate in the same group. If there are more than one part type that are selected by a priority rule, then we apply other priority rule to select a part type. We note that if the predicates *true* in the corresponding priority rule selects more than

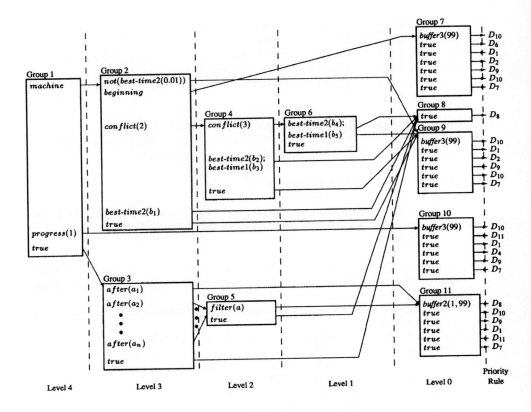

Figure 10.1. An example of the rule base.[4]

one part type then check their next predicate *true* until one part type is selected.

10.4 AN FMS SIMULATOR

An FMS is regarded as a dynamical discrete event system whose events are such as loading, processing, unloading, occurrence of machine failures, repairing of machine tools, and so forth. This section deals with an FMS with fixed routings so that sequences of machines to process are given for each job. Since stochastic Petri nets[5,22] are effective tools for modeling the dynamical and stochastic behavior of continuous-time discrete concurrent processes, they can be applied to model an FMS under uncertainty. In this article, we assume the followings for modeling FMS.

(1) Failures of machine tools or transports occur at random. The failure rates of machine tools or transports are constant, and the duration times that they will perform their intended function are random variables with exponential distribution with mean $1/\lambda$, where λ denotes failure rate.

(2) When machine tools or transports have failed, repair is started immediately.

(3) The time for repairing machine tools or transports is a random variable with exponential distribution.
(4) When machine tools or transports are finished to repair, they are operated immediately.
(5) Processing time is a random variable with normal distribution with mean value determined in process planning.

For describing hierarchical system such as FMS, in this article, we introduce *hierarchical timed transitions* which have a lower level subnet for calculating their firing times.

By using hierarchical stochastic Petri nets, we can model hierarchical systems that have the following characteristics.

(1) The whole system can be partitioned into submodels.
(2) The lower level submodel is regarded as a black box for the higher level submodel.
(3) The higher level submodel needs to receive the time from the lower level submodel when to start and when to finish execution.

Therefore, in modeling, we partition a model of an FMS into two parts; the processing level model and the transporting level model by using *hierarchical timed transitions,* and constructs the whole model by considering the processing level models as the submodels of the transporting level models. Furthermore, to reduce the computation time for simulation, we treat only the processing level models as continuous-time stochastic Petri nets and the transporting level models as discrete-time ones.

10.5 PERFORMANCE EVALUATION OF RULE BASE

Consider a hypothetical FMS which consists of the elements such as three machines *(a, b, c),* loading and unloading buffers for each machine, loading station and unloading stations, and a transport system with one automated guided vehicle. The system manufactures a family of products. The product routings are:

- Job A: $a(6) \rightarrow b(6)$,
- Job B: $a(4) \rightarrow b(3) \rightarrow c(5)$,
- Job C: $b(5) \rightarrow c(6)$.

Each number in the parentheses represents processing time units on a corresponding machine tool, where the number of each job is 20.

Comparing with some fixed priority rules shown in Table 10-1, we evaluate the rule base using the FMS simulator under the conditions shown in Table 10-2.

Figures 10.2 and 10.3 show the average completion time of 100 simulations and standard deviation of average completion time obtained by using the FMS simulator, respectively. All the average completion time obtained by using the rule base is much less than that obtained by using the fixed priority rules. Furthermore, almost all the standard deviation of completion time is less than that obtained by using the fixed priority rules. This implies that the rule base described in this article is superior to the other fixed priority rules.

Figure 10.4 shows average residual of completion time minus due date. If these values are nearly equal zero, it shows JIT is achieved. In Figure 10.4, all the value obtained by using

Table 10-1. Dispatching rules.

Rule base	Rule base
FCFS	First Come First Served
SPT	Shortest Processing Time
MINSLACK	Minimum Slack Time
MAXROB	Maximum Rate of Occupation for Buffers
MINROBN	Minimum Rate of Occupation for Buffers in Next Process
MAXLOSS	Maximum Sum of Loss Time
MINSPARE	Minimum Sum of Spare Time
MAXTARD	Maximum Tardiness

Table 10-2. Conditions of numerical experiments.

	Failure Rate	Variance of Processing Time	Mean Repair Time
Case 1	0.000	1.00	10
Case 2	0.001	1.00	10
Case 3	0.005	1.00	10
Case 4	0.010	1.00	10
Case 5	0.015	1.00	10
Case 6	0.005	0.00	10
Case 7	0.005	1.00	10
Case 8	0.005	2.00	10
Case 9	0.005	3.00	10
Case 10	0.005	1.00	5
Case 11	0.005	1.00	10
Case 12	0.005	1.00	15
Case 13	0.005	1.00	20

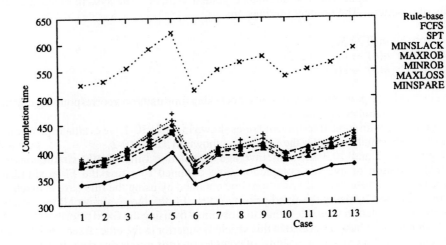

Figure 10.2. Completion time.

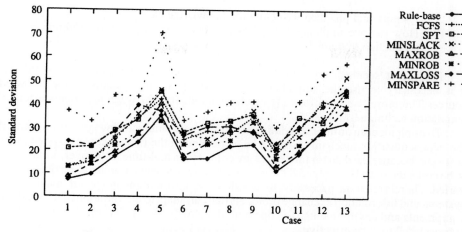

Figure 10.3. Standard deviation of completion time.

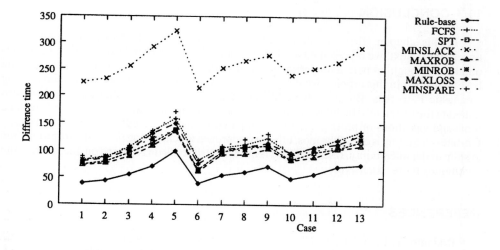

Figure 10.4. Average residual of completion time minus due date.

rule base is smaller than those obtained by using the other fixed priority rules. This concludes JIT is achieved by using rule base better than by using the others priority rules.

10.6 FUTURE STUDIES OF AI BASED ON-LINE SCHEDULING

To apply AI based on-line scheduling techniques to real production systems, it is necessary to resolve the following two problems:

(1) How can we obtain knowledges to develop a knowledge base in real production

systems and how can we debug the knowledge base?
(2) How can we implement the data acquisition mechanism for inferences in real production systems?

For (1), it seems that machine learning techniques[23] can be applied, but in this case, it is necessary to use the simulators of the manufacturing systems as the knowledge sources. Therefore, we need to study on the combination between machine learning and simulation techniques.

To implement an on-line scheduling system in real manufacturing systems, it seems that the data acquisition mechanisms must be embedded to the manufacturing system control software, because the data for inferences are collected in real-time. Moreover, kinds of data to be collected must be changed if the production environments or scheduling objectives are varied. Therefore, it is necessary to develop the information model of manufacturing systems and information exchange protocols that can adapt to the variation of the production equipments and environments. To cope with these difficulties, it seems that distributed AI approach[24, 25] may be effective.

10.7 CONCLUSION

In this chapter, first, we described a brief survey of scheduling problems under uncertainty. Next, we represented a rule base for on-line scheduling based on the heuristics obtained through numerical experiments. Furthermore, to construct and evaluate the rule base described in this article, we developed an FMS simulator under uncertainty using stochastic Petri nets. By using the FMS simulator, we confirmed that the rule base can generate the efficient schedules dynamically under uncertainty, such as failures of machine tools and variations of processing time, compared with other fixed typical priority rules. On-line scheduling is one of the most effective approach to scheduling problems for production system under uncertainty. It seems that it is necessary to study on AI based on-line scheduling techniques more extensively.

REFERENCES

1. Rodammer, F. and K. J. White. "A recent survey of production scheduling." *IEEE Transactions on Systems, Man, and Cybernetics.* Vol. 18, No. 6 (1988), pp. 841-851.

2. Baker, K. *Introduction to Sequencing and Scheduling.* John Wiley and Sons, New York (1974).

3. Harmonosky, C. and S. Robohn. "Real-time scheduling in computer integrated manufacturing: Review of recent research." *International Journal of Computer Integrated Manufacturing.* Vol. 4, No. 6 (1991), pp. 331-340.

4. Hatono, I., K. Yamagata, and H. Tamura. "Modeling and on-line scheduling of flexible manufacturing sytems using stochastic Petri nets." *IEEE Transactions on Software Engineering.* Vol. 17, No. 2 (1991), pp. 126-131.

5. Marsan, M., G. Conte, and G. Balbo. "A class of generalized stochastic Petri nets for the performance evaluation of mutiprocessor systems." *ACM Transactions on Computer System.* Vol. 2, No. 2 (1984), pp. 93-122.

6. Al-jaar, R. and A. Desrochers. "Performance evaluation of automated manufacturing systems using generalized stochastic Petri nets." *IEEE Transactions on Robotics and Automation.* Vol. 6, No. 6 (1990), pp. 621-639.

7. Carrie, A. *Simulation of Manufacturing Systems.* John Wiley and Sons, New York, 1988.

8. Fox, M. and S. Simith. "ISIS—a knowledge-based system for factory scheduling." *Expert Systems.* Vol. 1, No. 1 (1984), pp. 25-49.

9. Kanet, J. "Expert system in production scheduling." *European Journal of Operational Research.* Vol. 29, No. 1, 1984, pp. 51-59.

10. Fujii, S., Y. Watanabe, and H. Sandoh. "Optimal rescheduling policy based on real time disturbance information" *Proceedings of Japan-USA Symposium on Flexible Automation.* Kyoto (1990), pp. 1147-1150.

11. Agnetis, A., C. Aribib, M. Lucertini, and F. Nicolo. "Part routing in flexible assembly systems." *IEEE Transactions on Robotics and Automation.* Vol. 6, No. 6 (1990), pp. 697-705.

12. Kumar, P. and T. Seidman. "Dynamic instabilities and stabilization methods in distributed real-time scheduling of manufacturing systems." *IEEE Transactions on Automatic Control.* Vol. 35, No. 3 (1990), pp. 289-298.

13. Akella, R., B. Krogh, and M. Singh. "Efficient computation of coordinating controls in hierarchical structures for failure-prone multi-cell flexible assembly systems." *IEEE Transactions on Robotics and Automation.* Vol. 6, No. 6 (1990), pp. 659-671.

14. Brandimarte, B., C. Greco, and G. Menga. "Hybrid hierarchical scheduling and control systems in manufacturing." *IEEE Transactions on Robotics and Automation.* Vol. 6, No. 6 (1990), pp. 673-686.

15. Sharit, J. and G. Salvendy. "A real-time interactive computer model of a flexible manufacturing." *IIE Transactions.* Vol. 19, No. 2 (1987), pp. 167-177.

16. Ben-Arieh, D. and C. Moodie. "Knowledge based routing and sequencing for discrete part production." *Journal of Manufacturing Systems.* Vol. 6, No. 4 (1989), pp. 287-307.

17. Ranky, P. "Intelligent planning and dynamic scheduling of flexible manufacturing cells and systems." *Proceedings of Japan-USA Symposium on Flexible Automation.* San Francisco (1992), pp. 415-422.

18. Watanabe, T., H. Tokumaru, and Y. Hashimoto. "An analysis of dynamic scheduling for job-shop scheduling and its improvement using intelligent algorithms." *Proceedings of the IMACS/SICE International Symposium on Robotics, Mechatronics and Manufacturing Systems*. Kobe (1992), pp. 919-924.

19. Nakamura, Y., I. Hatono, Y. Kohara, and H. Tamura. "FMS scheduling using timed Petri net and rule base." *Proceedings of USA-Japan Symposium on Flexible Automation*. Minneapolis (1988), pp. 883-890.

20. Blackstone, J., D. Phillips, and G. Hogg. "A state-of-the-art survey of dispatching rules for manufacturing job shop operations." *International Journal of Production Research*. Vol. 20, No. 1 (1982), pp. 27-45.

21. Panwalker, S. and W. Iskander. "A survey of scheduling rules." *Operations Research*. Vol. 25, No. 1 (1977), pp. 45-61.

22. Archetti F. and A. Sciomachen. "Repesentation, analysis and simulation of manufacturing systems by Petri net based models." P. Varaiya and A. Kurzhanski eds., *Discrete Event Systems: Models and Applications*. Springer-Verlag, Berlin (1987), pp. 162-178.

23. Kodratoff, Y. and R. Michalski. *Machine Learning: An Artificial Intelligence Approach*. Volume III, Morgan Kaufmann, San Mateo (1990).

24. Parunak, H. "Distributed AI and manufacturing control: Some isuues and insights." In Y. Demazeau, et al. eds., *Decentralized A.I.* North-Holland, Amsterdam (1990), pp. 81-101.

25. Parunak, H. "Manufacturing experience with the contract net." In M. Huhns ed., *Distributed Artificial Intelligence*. Pitman, London (1987), pp. 923-928.

11

COORDINATION OF DISTRIBUTED INTELLIGENT SYSTEMS

M. Kamel and H. Ghenniwa
University of Waterloo
Waterloo, CANADA

Recently Distributed Artificial Intelligence (DAI) has become the main concern of many researchers in artificial intelligence. As a new field, DAI lacks theoretical foundations and a full understanding of its many aspects and problems. In particular, the problem of coordination which is central to DAI needs addressing. The objective of the research presented here is an attempt to develop a coherent body of coordination theory for distributed intelligent systems. This theory can be used to analyze when, which, and why a coordination technique is appropriate. Also it can be used as a tool for designing good coordination.

In this chapter, a brief literature review of theories, concepts and results from different disciplines that can contribute to the work of understanding and formulating the concept of coordination is overviewed, with special focus on DAI systems and organization theory. Then a 'Theory of Coordination' is developed in order to provide a quantitative representation of its main components and parameters, as well as performance measures.

A formulation of multi-agent systems (MAS) is developed to represent and develop the appropriate mechanisms that are required to handle coordination parameters. Finally a real-world domain, for assembling a table, is used to demonstrate the proposed theory and solution approaches.

11.1 INTRODUCTION

The word **distributed** is defined as the dispersion of functions and powers from a central authority to regional or local authorities. In this sense, a distributed system (DS) is constructed from several entities in which each independently is able to perform some functions and has to some extent the authority and the power to control its own functions. These entities can communicate together to work in the same special-time domain toward either a common goal or separate goals. Although there are many types of distributed systems, this work will focus on distributed systems that exhibit some sort of 'intelligence'. Distributed intelligent systems (DISs) consist of intelligent autonomous entities (agents) which are able to reason and act appropriately with respect to their environment such as done by a working group of human beings, an army of ants, or a group of intelligent robots. Further, the specific focus of this chapter is directed toward distributed artificial intelligence (DAI). From a survey of the most important work done in DAI research[1,11] it is reasonable to conclude that many important issues and problems are still not fully addressed. In particular, the coordination problem is not well understood and has been seen from different points of view by various researchers. This diversity of perspectives is largely due to the difficulty of the problem and the number of different considerations that are involved in solving it, such as problem-decomposition, problem-allocation, uncertainty, and interdependency. However, all of us may have an intuitive sense of what the word 'coordination' means. For instance, when we visit a smoothly functioning factory we may notice how well coordinated the actions of its departments and production lines seem to be. Usually, **good** coordination is plainly visible, however, lack of coordination can be more strikingly evident in a situation. For example, when two airplanes crash because of traffic instruction confusion we may become aware of the effect of poor coordination.

To have a formal understanding of the concept of coordination, it is helpful to have a more precise idea of what the term coordination means. Researchers from different disciplines have suggested various definitions for coordination such as: "the problem of arranging the signaling system for interdependent conditional activities";[27] "a distributed search through a space of possible interacting behaviors of individual agents and groups of agents to find a collection of behavior that satisfactorily achieves the agents' most important goals";[8] and "the act of managing interdependencies between activities".[26] The diversity of these definitions emphasizes the difficulty of defining **coordination**, as well as indicates the variety of possible starting points for studying the concept. The latter definition by Malone is the most appropriate for the work proposed here, and leads to the question: what does it mean **to manage**? This open question turns out to be crucial for several issues in the body of the coordination theory.

Presently, formal techniques of coordination in DAI such as the work of Rosenchein and his colleagues on formulating "Cooperation without Communication"[28] and "Probabilistic Interaction"[30] are beset by many weak and highly constraining assumptions. As well there is an experimental trend of developing different techniques for coordination such as "Meta-level Control",[2] "Partial Global Planning",[6] and its extensions based on Michigan

Intelligent Coordination Experiment (MICE)[8] which all suffer from being specialized and domain-specific. Despite the strong constraints that are attached to these techniques, it is important to realize that most of them have identified coordination techniques but did not provide a formal analysis of when, which, and why they are appropriate.

Consequently, the proposed work is directed toward developing a formal analysis and a quantitative measure of the coordination complexity. This measure then can be used to identify how this complexity can be reduced, and to analyze when, which, and why a coordination technique is appropriate.

The rest of the chapter is organized as follows. Section 11.2 reviews some of the tremendous amount of work that is related to the concept of coordination from a variety of research fields with special attention to DAI and organization theory. Section 11.3 sets the basis for a coherent body of coordination theory by providing formal and concrete definitions for its components and parameters. Section 11.4 provides an appropriate formulation for a DAI environment (Multi-Agent Systems MASs) to better specification of the coordination problem. A formulation is appropriate in the sense that it can be easily used to characterize the main components and parameters of the coordination problem, and helps to develop appropriate tools to handle each of them. Section 11.5 details a real-world example of a coordinated MAS. Finally, Section 11.6 concludes the main issues and aspects of the proposed research and summarizes its contribution.

11.2 RELATED WORK

The objective of this section is to summarize and stimulate development of theories that can help the proposed work to provide a precise understanding of the concept of coordination. A concrete definition of **coordination**, in turn, will help in identifying its main components and parameters of coordination.

In many different disciplines researchers have been concerned with how to coordinate agents' activities in distributed systems such as manufacturing machines, parallel and distributed computer systems (conventional and intelligent), and complex systems that include people and computers. However, some disciplines have restricted their definition or viewing of coordination to some aspects such as task allocation and conflict resolution. For example, distributed systems research considers the 'system-organization' metaphor in which the focus was to reduce the complexity of coordination.[9] Computer science restricts the issues of coordination to problem decomposition task allocation and interdependency concepts.[5,33] Operations research also focuses on the issue of optimality of decisions in task allocation and scheduling.[19,22] Economic theory has limits on its view of coordination to problem allocation.[35]

In the following subsections special attention is directed to coordination in DAI and human organizations.

11.2.1 Coordination in DAI

In DAI researchers have addressed many aspects of coordination,[1,11] however, solutions are focused on some specific problems. For example, contract-net protocol[4] treats coordination as task decomposition and distribution from a perspective in which there is one large problem and many potential agents, and the goal of coordination is to maximally utilize the agents.

Approaches based on FA/C[23] view coordination as the challenge of incompleteness, inconsistency, and incorrectness; or simply the uncertainty in control. Later, organizational structuring[2] has extended this view for better solution coherency by introducing the meta-level organization. Furthermore, partial-global-planing (PGP)[7] augmented organizational structuring for better solution coherency and convergence time. Agents in PGP exchange tentative plans information to help local plans to improve coordination among agents.

Social-community[12, 13] approach also views coordination as a process of organization based on settling and unsettling sets of questions. The open-system[16, 17] approach which is based on the scientific-community metaphor[21] is somewhat similar, however, it augments the social-community approach with the 'no-common' assumption.

Multi-agent planning looks to develop synchronized concurrent plans that account for agents' actions and interactions in achieving specific goals. This can be achieved either through an omniscient agent,[14, 15] or by a distributed approach[18] in which negotiation can be used for coordination.[32] Research in mathematical formalisms define coordination as conflict resolution in which an agent rationally chooses a move to maximize its payoff.[29, 36]

In recent work by Durfee[8] coordination is reduced to a distributed search through a space of possible interaction between individual agents and groups of agents to find a collection of behaviors that satisfactorily achieves the agents important goals. In Durfee's theory a metric system for coordination which consists of primary and secondary measures has been proposed. The primary is used to determine whether a potential resource conflict exists. The secondary is then used to measure the time at which the last activity of an agent concludes, in cases where no resource conflict has been found. Nevertheless, it is argued here that Durfee's theory can be viewed as a technique for 'Conflict-resolution'. Although, the secondary measure despite its restriction can be considered to some extent as a heuristic used to reduce the complexity of coordination.

11.2.2 Coordination in Human Organization

For human organizations March and Simon[27] defined the coordination problem as: "the problem of arranging a signaling system for 'interdependent' and 'conditional' (uncertain) activities". Interdependency implies that there is something of interest shared between parts of the organization. From the perspective of interdependency three kinds of coordination techniques have been identified:[34] (1) coordination by standardization for pooled interdependence, (2) coordination by plan for sequential interdependence, (3) coordination by mutual adjustment for reciprocal interdependence. However, the Information Processing school considers that the coordination difficulties arise if activities of the task rest on uncertainty. Uncertainty has been defined as the difference between the amount of information required to perform the task and the amount of information already possessed by the organization. Three mechanisms have been proposed to reduce the effect of uncertainty on coordination:[10] (1) environmental management, (2) the creation of slack resources, (3) the creation of self-contained tasks. In addition to the interdependency and uncertainty, Simon[31] identified a very important factor that affects the coordination difficulties which is called **bounded-rationality**: the capacity of individual mind for formulating and solving complex problems is very small compared with the size of the problems whose solution is required for objectively rational behavior in the real world or even for a reasonable approximation to such objective rationality. To deal with bounded-rationality, two mechanisms have been proposed to increase the processing capacity of the agents:[10] (1) investment in vertical information systems, (2) creation of lateral relations.

Recently, a group at MIT established the Center for Coordination Science to study and develop an Interdisciplinary Theory of Coordination (Malone's theory). Although, their work is still in the exploratory phase[3, 24, 25, 26] some insights can be gained from their definition of coordination. Malone's theory is a set of principles about how activities can be coordinated. In their view coordination is based on four components: goals, activities, actors (agents), and interdependencies. Each of these components is associated with coordination processes; for instance, goals are associated with a coordination process called identifying goals (goal selection). In turn each of these processes requires underlying processes such as group decision making, communication, and perception of common objects. Although this theory offers a precise definition of the concept of coordination, it favors a comprehensive classification of strategies for coordination mechanisms rather than a theoretical formulation. Also it is argued here that agents, goals, and activities are more appropriately elements of the environment (may suffer from coordination problems) than of the coordination problem itself . The key issue of determining the exact components (structure and mechanisms) and parameters (informally, interdependency and uncertainty) of coordination is to understand what it means 'to manage the interdependency' in the definition of coordination. These components and parameters are truly the characteristics of the coordination itself, and not of the environment. In this sense, Malone's theory can be considered as a classification approach for different processes that can be used to handle coordination within the limits of the environmental components instead of identifying coordination components and parameters.

The research outlined here attempts to address some of the weaknesses that are exhibited by most of the existing work and to provide a more complete picture of coordination theory. Next, a coherent body of coordination theory is developed in order to provide a formal determination of its components and parameters, as well as a quantitative representation for its performance measure. This formulation is then used as a tool to determine a 'good' coordination. Also, a formulation of multi-agent systems (MAS) is developed to represent the coordination, in a way it helps to identify and develop the appropriate mechanisms that are required to handle coordination parameters.

11.3 TOWARDS A COORDINATION THEORY

To visualize the abstract perspective of coordination, consider the following example: a work-cell contains two agents, A and B. Each agent has its own exclusive workbench and all the resources it needs to build a certain product (product-1 for A and product-2 for B) except for two tools, tool-1 and tool-2, which both agents share. The final product which combines product-1 and -2 cannot be assembled by a single agent.

In the following subsections a precise and a complete definition of coordination is proposed (based on Malone's[26]), which is then used to identify and examine its main components and parameters. Then a quantitative measure for the complexity of coordination is developed. This measure can be used to analyze when, which, and why a coordination mechanism (device) is appropriate.

11.3.1 Coordination: A Concrete Definition

In a distributed environment such as the above **Assembly** scenario, agents have to coordinate their efforts to achieve their goal(s). For instance, agents A and B have to share their use of tool-1 and tool-2, although, they can work independently at their respective workbenches. Nevertheless, they have to meet in a shared work-space in order to combine their products.

In this work coordination is the act of managing interdependencies between activities performed by agents to achieve some goal(s). **Activities** are physical or mental efforts such as the move action or the process of information respectively. **Goals** are the ends that activities are directed toward such as 'product-1 at shared-floor'. **Interdependencies** are goal-relevant relationships between activities. For example, agents A and B each has a goal that cannot be achieved without using tool-1, and consequently this imposes some sort of interdependence between the agents' activities. Finally, **to manage** means to provide coordination **structure** and **mechanism** which constitute the two main components of coordination.

Coordination structure is the pattern of decision making and communication among a set of agents who perform tasks in order to achieve goals. Two basic types of coordination structure can be identified, centralized and decentralized. In the centralized, shown in Figure 11.1, only one agent is allowed to be a decision maker or a coordinator, whereas, in the decentralized, shown in Figure 11.2, each agent is allowed to contribute to the coordination.

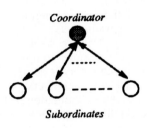

Figure 11.1. A centralized coordination structure.

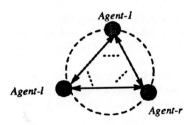

Figure 11.2. A decentralized coordination structure.

In many cases, however, problems have a considerably large size and complexity to the extend that the limits of agents' capabilities are reached. Solving such problems using the basic structures become very costly or perhaps impossible to solve. Nevertheless, it is easy to build on these basic types of structure to fit the problem characteristics of size and complexity with respect to the agents' bounded-capability. Firstly, the number of agents have to be increased in order to reduce the bounded-capability effect. For example, if two agents aren't enough to build a tower starting from the designing stage down to painting the walls a much larger number of agents is required, although, two agents might be enough to paint a wall.

Secondly, dividing the coordination structure into levels (**multi-level hierarchy**) that are equivalent to the levels of the problem abstractions becomes a natural extension. Subproblems at each level of abstraction can be solved only within its knowledge domain. Bounded-capability effect then can be diminished by reducing the complexity of the problems at each level by decomposing a large and (or) complex problem to smaller and less complex subproblems. For example, 'build a tower' can be considered as a problem at the highest level of the abstraction, with 'paint a wall' a problem to be solved at some lower level. In this scenario a large number of agents are involved, starting from the owner-agent at the highest level of the abstraction down to the painting-agent at a lower level in which the owner-agent is not directly involved. Activities of both agents and all the others are coordinated through a multi-level hierarchical structure. In a hierarchical structure each agent is subject to coordinator-subordinate relationships. Each agent's coordination problems except for the top level agent(s) are entitled to be solved by its coordinator. In this way no activity is left uncoordinated. Thus, this kind of coordination structure, to some extent, is centralized.

Finally, as the number of agents increases due to problem size and complexity, multi-agent systems take the form of a **multi-organization** society, where a group of units interact at some level of problem abstraction. An organization unit is a group of agents that has its own intra-coordination structure. For a multi-organization society **market** is an appropriate coordination structure, where a number of disjoint organization units are assumed to be available for providing a service. Activities are initiated after a successful negotiation of a contract between the units. By which, the size and the complexity of some of the problems can be reduced to messages of contract negotiation. For example, consider a factory that requires many different parts such as raw material, glass, and door knobs in order to manufacture furniture. As an organization unit, this factory reduces some of the problem's size and complexity to a few messages by contracting with some other units to provide raw-material, glass, and door knobs. In this sense, market is a decentralized structure at some level of problem abstraction.

Structure as the first main component of coordination ranges from strict one level to heterarchical structures, depending on the problem's size, complexity, and the number of agents. The level of coordination structure is corresponds to the abstraction level of the problem. However, the type of coordination structure, in general, remains centralized or decentralized.

Coordination mechanism, as the second component of coordination, is a device or a collection of devices which can be used to handle the coordination parameters: informally, type of interdependence and uncertainty. In distributed systems three types of interdependence can be identified: (1) physical, such as interdependence imposed by sharing tool-1; (2) mental, for example an agent's activity depends on some knowledge produced by another agent; (3) time, such as a sequential temporal order between some agents' activities. Each type of interdependence may exhibit a different topology which affects the flow of coordination, and thus its difficulties. Although a different coordination technique is required for each interdependence type and topology, coordination difficulties arise due to the uncertainty of interdependence.

Hence, uncertainty, as the second parameter, is characterized in terms of: incorrect-ness which refers to the amount of error that the knowledge manifests, and incompleteness which refers to the lack of having the necessary requirements. Uncertainty is exhibited in three forms, task-knowledge, control, and environment. However, the main focus of this research is the incompleteness aspect of task-knowledge and control uncertainty and their impact on coordination.

Knowledge, here, is treated as a set of propositions, and it is available if it is presented explicitly or implicitly in the agent's memory. Task-knowledge is the knowledge that is required to perform a task. Hence, **incompleteness of task-knowledge** is the difference between the necessary knowledge to perform a task and the available knowledge. For example, let tool-1 in the Assembly scenario be a drill and assume that the necessary knowledge to make a hole is the location of the hole, its diameter, and the status of the drill. Given that an agent aiming to make a hole doesn't know its location, obviously, the incompleteness set of the task-knowledge will contain the location of the hole. A major problem with determining the incompleteness of task-knowledge arises when knowledge is implicitly available, or the omniscience problem.

Control, here, is treated as a set of decision points at which knowledge is required to determine the next activity in order to reach the desired end (goal). This control is exercised either by the agent or for the agent. Two types of control can be identified for a distributed environment local-control and coordinated-control. Local-control is a set of decision points which determine the cheapest sound set of activities (the optimal solution) or the closest set to the optimal. Nevertheless, this research is concerned with coordinated-control. **Coordinated control** is a set of decision points which represent the influence of the existence of the other agents on the local-control. This influence is caused by interdependency and coherency. It is obvious that in distributed environments decision points have to consider which and when a physical interdependence is to be acquired or released, a mental interdependence is to be produced or consumed, and activities should act in sequential or synchronized fashion. This is called interdependency-knowledge. For example, assume that tool-1 and tool-2 are equivalent and the status of tool-1 is busy. A decision point of (or for) an agent aiming to make a hole is affected by the knowledge of the tools' status the agent has to decide whether to wait for tool-1 to be free or to use tool-2. **Incompleteness of interdependency-knowledge** is the difference between the shared resources, the other agents' capabilities as well as their goals, and the available knowledge. In spite of this, the selected activities which are locally the best should also be globally the most beneficial. In terms of this solution-coherency, **incompleteness of coherency-knowledge** is the difference between the other agents' goals as well as their solutions, and the available knowledge. From this perspective, the main objective of a coordinated-control is to provide the local control with the ability to determine which alternative set of activities requires less interdependence and leads to solutions that are globally the most beneficial.

By this a concrete definition of **coordination** is completed, next a quantitative representation for the coordination performance (or complexity) is proposed.

11.3.2 Performance Measure of Coordination

Based on the above definitions, in the following subsections, a mathematical measure of coordination performance is formulated in terms of both structure and mechanism complexities. By complexity, here, we mean the cost of the time-span and computational efforts. The structure-complexity is formulated on the basis of computational and communication costs required for coordination. This measure is used to decide on the appropriate structure that minimizes the complexity. The formulation of the coordination mechanism is based on two objectives: 'reduce' and 'resolve' coordination problems associated with each parameter. The mechanism-complexity is, therefore, formulated using a function that measures the quality of heuristics used to reduce the effect of each parameter, and a function that measures the efficiency of devices used to resolve coordination problems associated with each parameter. By these measures it is easy to analyze when, which, and why a coordination technique is appropriate, and to design a good coordination.

A Quantitative Representation of the Parameters Let P^l be a problem to be solved at abstraction level-l. For a given set of agents characterized by limited capabilities, the effect of the system's bounded-capability can be reduced by breaking down the original problem into smaller sub-problems. Assume that P^l, a decomposition device at abstraction level-l, decomposes p^l into a set of subproblems (goals) denoted by $G^l = \{g_1, g_2, ..., g_{\lambda_1}\}$, where λ_1 represents the problem size. A goal is considered as a set of conditions that the world should satisfy. A goal is called persistent if it is part of the final goal, and called intermediate if it should hold until some other goals are satisfied. Five types of relationships between goals can be identified:

1. g_i and g_j are said to be **contradicting** goals, and denoted by $g_i \Updownarrow g_j$, if g_i and g_j are persistent and for any possible world that satisfies g_i, it is impossible for g_j to be satisfied;
2. g_i and g_j are said to be **conflicting** goals, and denoted by $g_i \diamondsuit g_j$, if
   a) g_i and g_j are persistent and for some worlds that satisfy g_i, it is possible to change this world to another world that satisfies g_j with a tentative disturbance to g_i;
   b) g_i is intermediate and for any possible world that satisfies g_i it is impossible for g_j to be satisfied without disturbing g_i;
3. g_i and g_j are said to be **common** goals, and denoted by $g_i \equiv g_j$, if for any possible world that satisfies g_i it also satisfies g_j.
4. g_i and g_j are said to be **independent** goals, and denoted by $g_i \| g_j$, if for any possible world that satisfies g_i, there is always a way to change this world without disturbing g_i to another world that satisfies g_j, and vice versa;
5. g_i and g_j are said to be **simultaneous** goals, and denoted by $g_i \triangleright\triangleleft g_j$, if they are independent goals, and for all worlds that are selected to satisfy gi, g_j should also be satisfied.

It is assumed that it is impossible to solve a problem that contains contradicting goals. Therefore, the main focus is how to solve problems that arise due to common, conflicting, independent, and simultaneous goals. Note, goals with independent relationships don't require coordination. Based on the above definitions, for a problem with a size λ_1, there are $C(2, \lambda_1) = \dfrac{\lambda_1(\lambda_1 - 1)}{2}$ relationships between goals. Let *Nf*, *Nc*, and *Ns* be the number of conflicting, common, and simultaneous goals respectively; $P_f, P_f,$ and P_f be the probability that an agent may be assigned a conflicting, a common, or a simultaneous goal respectively, where $P_f = \dfrac{2Nf}{\lambda_1(\lambda_1-1)}$, $P_c = \dfrac{2Nc}{\lambda_1(\lambda_1-1)}$, and $P_s = \dfrac{2Ns}{\lambda_1(\lambda_1-1)}$. Now, assume a multi-agent system consists of m agents is used to solve G^l. Let:

* Ac_i be the set of activities that agent i can perform;

* Pr_{ij} be the set of preconditions of activity j of agent i;

* D_i be the set of preconditions of agent i's activities that are not in the effect set of its activities;

* R_i be the set of agent i's activities which require resources;

* Kg_i be the available knowledge of agent i.

Therefore, the probability that:

* agent i is not capable of solving a problem is denoted by $\overline{P_{p_i}} = 1 - \dfrac{|Ac_i|}{|\bigcup_{j=1}^{m} Ac_j|}$, for simplicity assume $\overline{P_{p_i}} = \overline{P_p} = (1 - P_p)$;

* agent i is a dependent agent is denoted by $P_{d_i} = \dfrac{|D_i|}{|\bigcup_{j=1}^{l} Pr_{ij}|}$, for simplicity assume $P_{d_i} = P_d = (1 - \overline{P_d})$;

* agent i may require a resource is denoted by $P_{r_i} = \dfrac{|R_i|}{|Ac_i|}$ for simplicity assume $P_{r_i} = P_r$;

* agent i does not know some knowledge is denoted by $\overline{P_{k_i}} = \dfrac{|kg(Ac_i,t) - Kg_i|}{|kg(Ac_i,t)|}$, where $kg(Ac_i,t)$ is a function which returns a set of all knowledge that is required for all activities of agent i at a specific instance of time, t, for simplicity assume $\overline{P_{k_i}} = \overline{P_k} = (1 - P_k)$; .

Coordination of Distributed Intelligent Systems

For given P_f, P_c, and P_s: the probability that an agent has a goal having a conflicting, common, or simultaneous relationship with the others respectively are $\frac{P_f}{m}$, $\frac{P_c}{m}$, and $\frac{P_s}{m}$. Furthermore, let

* cc be the computation costs—computation time required to perform a coordination activity;
* cm be the communication costs—communication time required to perform a coordination activity;
* cd be delay cost associated with a single coordination activity such that performing a specific activity before another imposes a time delay.

For generality, assume that subproblems are randomly assigned to agents. This assumption allows the main parameters that affect coordination to be pointed out. Also, assume that the load balance is achievable such that an assignment device Λ^l assigns $\frac{\lambda_1}{m}$ subproblems to each agent in a society of m agents. Furthermore, assume that P^l can be solved after rounds of k coordination, and for each round solving a subproblem requires λ'_k subproblems $1 \leq k' \leq k$.

Thus, the incompleteness of task-knowledge can be quantitatively represented by:

* the average number of subproblems that cannot be solved by agent i which is $\overline{P_p} \frac{\lambda_1}{m}$;

* the average number of activities that require knowledge which is

$$\overline{P_k}\, \overline{P_d}\, P_p\, \frac{\lambda_1}{m}\, \lambda'_k\ .$$

The incompleteness of interdependency-knowledge can be represented in terms of:

* the average number of dependent activities which is $P_d\, P_p\, \frac{\lambda_1}{m}\, \lambda'_k$;

* the average number of activities that require resources which is $P_r\, \overline{P_d}\, P_p\, \frac{\lambda_1}{m}\, \lambda'_k$;

* the average number of conflicting goals which is $\frac{P_f}{m} \overline{P_d} P_p \frac{\lambda_1}{m} \lambda_k^{'}$;

* the average number of common goals which is $\frac{P_c}{m} \overline{P_d} P_p \frac{\lambda_1}{m} \lambda_k^{'}$;

* the average number of simultaneous goals which is $\frac{P_s}{m} \overline{P_d} P_p \frac{\lambda_1}{m} \lambda_k^{'}$.

Also, the incompleteness of coherency-knowledge can be represented as:

* the average number of conflicting goals which is $\frac{P_f}{m} \overline{P_d} P_p \frac{\lambda_1}{m} \lambda_k^{'}$;

* the average number of common goals which is $\frac{P_c}{m} \overline{P_d} P_p \frac{\lambda_1}{m}$;

* the average number of simultaneous goals which is $\frac{P_s}{m} \overline{P_d} P_p \frac{\lambda_1}{m} \lambda_k^{'}$.

Now, the coordination parameters will be referred by:

$$\Gamma_p = \{P_p, P_d, P_k, P_r, P_f, P_c, P_s, \lambda_1, \lambda^{'}\}.$$

The Structure Complexity The coordination-complexity imposed by the structure is formally given by $Sc^v = Cmp^v + Com^v$, where

* $v \in \{c,d\}$, c for centralized, and d for decentralized coordination structure;
* Cmp^v denotes the computation and delay costs for processing the coordination activities, for simplicity it is called the processing cost;
* Com^v is the computation and delay costs for communication required for coordination, for simplicity it is called the communication cost.

In the following two subsections structure-complexity, Sc^v, is determined for both centralized and decentralized structures in terms of the coordination parameters. Note that by definition it is easy to adopt these formulations to accommodate different structure levels such as multi-level hierarchical or market.

Centralized Structure Consider a centralized structure with a coordinator agent Ag_c, and m subordinate agents denoted by $\{Ag_1, Ag_2, ..., Ag_m\}$. According to the definition of the centralized structure, solving all of the coordination problems is the responsibility of Ag_c. However, performing the coordination activities is the responsibility of all agents. For example, Ag_i finding out that it needs help from another agent is a coordination activity, but deciding which agent is able to help Ag_i is the responsibility of Ag_c. The processing cost of Ag_c is denoted by cmp_c^c, and that of Ag_i by cmp_{si}^c where $cmp_s^c = \sum_{i=1}^m cmp_{si}^c$. Hence, $Cmp^c = cmp_c^c + cmp_s^c$.

In the first round of coordination Ag_c performs computations for coordination due to $\Lambda^l(P^l)$ with cost given by: $cmp_c^c = \lambda_1(cc + cd)$. Each of its subordinates also performs coordination activities with processing cost given by:

$$cmp_{si}^c = \frac{\lambda_1}{m}\left[\overline{P_p} + \lambda_1 P_p\left(P_d + \overline{P_d}\left(\overline{P_k} + P_r + \frac{1}{m}(P_f + P_c + P_s)\right)\right)\right](cc + cd).$$

In the second round Ag_c performs $\Lambda^l(\lambda_2)$, where $\lambda_2 = \lambda_1(\overline{P_p} + P_d P_p \lambda_1)$. In addition to this it solves problems for task-knowledge and coordinated-control (interdependency and coherency). Thus, the computational cost for Ag_c is given by:

$$cmp_c^c = m\frac{\lambda_1}{m}\left[\overline{P_p} + \lambda_1 P_p\left(P_d + \overline{P_d}\left(\overline{P_k} + P_r + \frac{1}{m}(P_f + P_c + P_s)\right)\right)\right](cc + cd); \text{ and}$$

the processing cost for each subordinate is given by:

$$cmp_{si}^c = \frac{\lambda_2}{m}\left[\overline{P_p} + \lambda_2 P_p\left(P_d + \overline{P_d}\left(\overline{P_k} + P_r + \frac{1}{m}(P_f + P_c + P_s)\right)\right)\right](cc + cd).$$

Finally, in the kth round:

$$cmp_c^c = \lambda_1 \sum_{\kappa=2}^{k}\left(\left\lfloor\frac{2}{\kappa}\right\rfloor + \prod_{i=1}^{\kappa-2}(\overline{P_p} + P_d P_p \lambda_i)\left[\overline{P_p} + \lambda_{\kappa-1} P_p\left(P_d + \overline{P_d}\left(\overline{P_k} + P_r + \frac{1}{m}(P_f + P_c + P_s)\right)\right)\right]\right)(cc + cd).$$

$$cmp_s^c = \lambda_1 \sum_{\kappa=1}^{k-1}\left(\left\lfloor\frac{1}{\kappa}\right\rfloor + \prod_{i=1}^{\kappa-1}(\overline{P_p} + P_d P_p \lambda'_i)\left[\overline{P_p} + \lambda_2 P_p\left(P_d + \overline{P_d}\left(\overline{P_k} + P_r + \frac{1}{m}(P_f + P_c + P_s)\right)\right)\right]\right)(cc + cd).$$

Let Z be the total coordination activities, hence, $\dfrac{Z}{m} = \dfrac{cmp_s^c}{m(cc) + cd}$. Therefore,

$$Cmp^c = (Z+\lambda_1)(cc+cd) + Z\left(cc+\frac{cd}{m}\right);$$

$$= (2Z+\lambda_1)cc + \left(\frac{m+1}{m}Z+\lambda_1\right)cd.$$

Obviously,

$$Com^c = (Z+\lambda_1)(cm+cd) + Z(cc+cd),$$

$$= (2Z+\lambda_1)cm + (2Z+\lambda_1)cd.$$

Therefore,

$$Sc^c = (2Z+\lambda_1)cc + (2Z+\lambda_1)cm + \left(\frac{3m+1}{m}Z+2\lambda_1\right)cd.$$

Decentralized Structure Consider a decentralized structure with n agents denoted by $\{Ag_1, Ag_2, ..., Ag_n\}$, where solving coordination problems is the responsibility of every agent. During every round each agent acts as a coordinator and the others as subordinates. For example, Ag_i is responsible for finding out if it needs help from another agent, determining which agents are able to help it, and delegating these activities to the others. The processing cost of Ag_i for coordination is denoted by cmp_i^d, where $1 \le i \le n$.

In the first round it is assumed that each agent has to perform $\Lambda'(P')$, therefore, $cmp_i^d = \lambda_1(cc+cd)$. In the second round it is assumed that each agent is assigned $\lambda_2 = \frac{\lambda_1}{n}$ problems, therefore,

$$cmp_i^d = \frac{\lambda_1}{n}\left[\lambda_2'\left(P_d + \overline{P_d}\left(\overline{P_k} + P_r + \frac{1}{n}(P_f + P_c + P_s)\right)\right)\right](cc+cd).$$

Finally, in the kth round:

$$cmp_i^d = \lambda_1 \begin{pmatrix} 1+\frac{1}{n}\left(P_d + \overline{P_d}\left(\overline{P_k} + P_r + \frac{1}{n}(P_f + P_c + P_s)\right)\right)+ \\ \sum_{\kappa=3}^{k-1} P_d \lambda_1 \frac{\lambda_2'}{n}\left(\left\lfloor\frac{3}{\kappa}\right\rfloor + \prod_{t=3}^{\kappa}(\overline{P_p} + P_d P_p \lambda_1)\right)\left[\overline{P_p} + \lambda_\kappa' P_p\left(P_d + \overline{P_d}\left(\overline{P_k} + P_r + \frac{1}{n}(P_f + P_c + P_s)\right)\right)\right] \end{pmatrix}(cc+cd).$$

This can be rewritten as

$$cmp_i^d = \left(\frac{Z}{n} + \lambda_1\left(1 - \frac{\overline{P_p}}{n}\right)\right)(cc + cd).$$

Therefore, $\quad Cmp^d = \left(\frac{Z}{n} + \lambda_1\left(1 - \frac{\overline{P_p}}{n}\right)\right)(n \times cc + cd).$

Obviously, $\quad Com^d = \left(\frac{Z}{n} + \lambda_1\left(1 - \frac{\overline{P_p}}{n}\right)\right)(n(n-1)cm + (n-1)cd).$

Therefore,

$$Sc^d = \left(Z + n\lambda_1\left(1 - \frac{\overline{P_p}}{n}\right)\right)cc + \left(Z + n\lambda_1\left(1 - \frac{\overline{P_p}}{n}\right)\right)(n-1)cm + 2n\left(\frac{Z}{n} + \lambda_1\left(1 - \frac{\overline{P_p}}{n}\right)\right)cd.$$

The Mechanism Complexity Let Ξ be a coordination mechanism applied to handle each parameter of Z. The main objectives of Ξ are to **reduce** and **resolve** coordination problems associated with each parameter. Reducing coordination problems can be formally stated as *minimizing Z*. Assume that x_ρ is a variable introduced to control the effect of each parameter in Z, where $0 \le x_\rho \le 1$, $\rho \in \{p, d, k, r, f, c, s, o\}$. Since the parameters of Z are independent of each other, then *minimizing Z* is equivalent to: $P_\rho = minimize\ (x_\rho P_\rho)$, for $\rho \in \{d, r, f, c, s, o\}$, and $\overline{P_{\rho'}} = minimize\ (x_\rho \overline{P_\rho})$, for $\rho \in \{p, k\}$. To understand the precise meaning of x_ρ, assume that different heuristics (such as those in Table 1) are available to be used for minimizing the effect of some parameters of Z. Thus, for a heuristic H_ρ which can be used to minimize the effect of P_ρ, there is a corresponding $x_\rho = \Theta(H_\rho)$, where Θ is a function which measures the quality of H_ρ. Hence, x_ρ is equal to zero when H_ρ is perfect and equal to one when H_ρ is worse for P_ρ.

To resolve problems associated with each parameter there are different techniques that can be employed. For example, the conflict problem (due to P_j) can be resolved using techniques such as assigning authority levels among agents, consulting a mediator, or negotiation. Different techniques exhibit different computational cost. Thus, the complexity of a coordination mechanism, in general, can be measured by

Table 11-1. Some general heuristics for minizing the effect of Z's parameters.

Parameter	Heuristics
x_p:	* Assign problems to the capable agents.
x_d:	* Assign problems to the most 'self-contained' agents. * Avoid the selection of dependent activities.
x_k:	* Select activities that require less knowledge. * Get knowledge whenever it is possible.
x_r:	* Assign problems that share resources to the same group of agents. * Maintain 'load-balance' state for resources.
x_f:	* Assign conflict goals to the same group of agents. * Avoid generating conflict goals.
x_c:	* Assign common goals to the same group of agents. * Avoid generating redundant (common) goals.
x_o:	* Select the cheapest solution.

$$Mc = \lambda_1 \sum_{\kappa=1}^{k-1} \left(\prod_{t=1}^{\kappa-2} \left(\overline{P_{p'}} + P_{d'} P_{p'} \lambda_t^o \right) \left[\overline{P_{p'}} |D_p| + \lambda_{\kappa-1}^o P_{p'} \left(P_d |D_d| + \overline{P_{d'}} \left(\overline{P_{k'}} |D_k| + P_{r'} |D_r| + \frac{1}{m} \right. \right. \right. \right.$$
$$\left. \left. \left. \left. \left(P_{f'} |D_f| + P_{c'} |D_c| + P_{s'} |D_s| \right) \right) \right] \right) (cc + cd).$$

where $|D_\rho|, \rho \in \{p,d,k,r,f,s,c\}$, is the computational cost of the device ρ.

Thus, the optimality of coordination for a given structure can be measured by:

$$\min \sum_\rho x_\rho,$$
$$\min \sum_\rho |D_\rho|,$$
$$s.t.$$
$$0 \leq x_\rho \leq 1, and$$
$$|D_\rho| \geq 1.$$

The first objective function measures the quality of the heuristics used to reduce the coordination problems. The second measures the efficiency of the devices used to resolve the corresponding problems.

In the following sections, a formal representation of **coordination** in a DAI environment (Multi-Agent System MAS) is developed by direct transformation of the coordination components and parameters to their corresponding in the environment. This helps to determine the main specification of generic problem-solving devices for each parameter.

11.4 FORMULATION OF A DAI ENVIRONMENT

In this research, the main concern is solving a problem in a multi-agent setting in which planning is considered to be a problem-solving mechanism. A **multi-agent system** (MAS) is a collection of cooperative, autonomous, and intelligent agents that have the capability to coordinate with each other in order to solve some problem(s). These agents may be working toward a single global goal or toward individual goals with interdependent relationships. An **intelligent** agent is an entity that has the capability to act rationally and intentionally with respect to its own and the society's goals. An **autonomous** agent refers to an entity that has its own existence which is not justified by the existence of the other agents. Cooperative agents help each other on trading knowledge as well as in carrying out tasks that are requested from them. Formally, a multi-agent system is a 4-tuple, $MAS = \langle A, O, \Re, \Im \rangle$, where

1. a set of m cooperative, autonomous, and intelligent agents, denoted by
   $A = \{Ag_1, Ag_2, ..., Ag_m\}$;

2. a set of n objects that are manipulated (processed) by A, and denoted by
   $O = \{o_1, o_2, ..., o_n\}$;

3. a set of k resources that are required by A to process O, and denoted by
   $\Re = \{r_1, r_2, ..., r_k\}$;

4. a global time line given by $\Im = \langle T_d, T_c \rangle$, where
   * T_d is a set of time instances (discrete notion of time) denoted by
     $T_d = \{t_l | 1 \leq l \leq \alpha\}$;
   * T_c is a set of time intervals or periods (continuous notion of time)
     denoted by $T_c = \{\tau_j | 1 \leq j \leq \beta\}$

Elements of O, $\Re$, T_d and T_c are characterized by x, y, z_1, z_2-tuple relations which are respectively denoted by $Or^x, Rr^y, Tr_d^{z_1}, Tr_c^{z_2}$. An agent Ag_i is a 2-tuple $\langle Kg^i, Cp^i \rangle$, where Kg^i denotes Ag_i's knowledge and Cp^i denotes its capabilities.

* Kg_i is a 3-tuple, $\langle Pd^i, Kc^i, M^i \rangle$, where
  - Pd^i is a 5-tuple $\langle P, G^i, W_T^i, Pl_T^i, I_T^i \rangle$ denoting the problem domain specification, where P is a set of problems, G_T^i is a set of goals or subproblems, W_T^i denotes a local world history; Pl_T^i denotes Ag_i's plans for achieving P or G_T^i; and I_T^i denotes Ag_i intentions or persistence plans.
  - Kc^i is a procedural knowledge of each capability item.
  - M^i is a set of other agents' models denoted by $M^i = \{M_l^i | 1 \leq l \leq m, i \neq l\}$

and, $M_l^i = \langle X_1^i, X_2^i, \ldots, X_x^i \rangle$; where the definition of the X^is are application dependent.

* Ag_i 's capabilities, Cp^i, is a 3-tuple, $\langle Rs^i, Cm^i, Ac^i \rangle$, where
  - Rs^i denotes its reasoning capabilities such as reasoning about problem-domain and coordination;
  - CM^l is Ag_i 's communication capabilities such as communication protocol and bandwidth;
  - Ac^i is a set of domain-dependent activities belonging to Ag_i.

Planning approaches use various types of world representation. Here, the state-base, temporal world notion is proposed as a world representation for the problem-solving mechanism. A complete, consistent, and global snapshot of the world, which defines a world state is denoted by $s = \langle O, Or^x, \Re, Rr^y \rangle$. An incomplete version of s which represents a local view of the world state, or world situation, is denoted by $s' = \langle O', Or'^x, \Re', Rr'^y \rangle$, where $O' \subset O, Or'^x \subset Or^y, \Re' \subset \Re$, and $Rr'^n \subset Rr^n$. A global view of the world history is given by a function denoted by $W_T: T \to S$, where $S = \{s \mid s \text{ is a possible state}\}$. A local view of the world history is defined as a function denoted by $\omega_T: T \to S'$, where $S' = \{s' \mid s' \text{ is a possible situation}\}$. A change on a world state or situation results from applying domain-dependent activities. A domain-dependent activity a_j^i, read as activity j belongs to agent Ag_i, is defined as a function which maps an agent name Ag, a set of situations S, and a time interval τ into a set of situations S, and denoted by $a_j^i: Ag \times S \times \tau \to S$. Activities may be interrelated by 2-tuple relations, which can be determined by a function denoted by $Ac_r: \bigcup_{i=1}^m Ac^i \times \bigcup_{i=1}^m Ac^i \to \{\Rightarrow, \mapsto, \Leftrightarrow\}$, where $\Rightarrow$ is a temporal order (precedence) inter-relationship, $\mapsto$ is a causal inter-relationship (direct causal), and $\Leftrightarrow$ is a simultaneity inter-relationship.

Finally, to solve a problem p an agent must be able to reason about the problem domain. Thus, problem-solving (Ps) is a function that maps local world history, goals, domain-dependent activities, and the agent's intentions into a plan. That is, $Ps: \omega_T^i \times G^i \times Ac^i \times I_T^i \to Pl_T^i$. A plan is a time indexed (over T_d) list of actions given by $Pl_T^i = \{(a_j^i, t) \mid t \in T_d\}$. These plans are violable commitments in the sense that they should be granted by the society in order to be intentions. Plans can be brought to intentions by reasoning about coordination using a two step mechanism. The first step is called **semi-coordinated** problem-solving which is a function that maps plans, the agents' model, goals, and a set of resources into a coordinated plan, and denoted by $Sr: Pl_T^i \times M^i \times G^i \times \Re \to C_{pl_T^i}$. The second step is called **coordinated** problem-solving which is a function that maps the coordinated plans and the communication capabilities into intentions, and denoted by $Cr: C_{pl_T^i} \times Cm^i \to I_T^i$.

In the following subsections, the coordination theory, developed in the previous sections, is used to understand the precise function and to determine the main functions and parts of the semi-coordinated and the coordinated problem-solving in order to achieve 'good' coordination.

11.4.1 Coordination of DAI Systems

In this section the main components and parameters of coordination (developed in the previous sections) are transformed into a DAI environment using the above formulation. Formulating coordination structure is briefly discussed, and then a detailed formulation of coordination mechanism and solution approaches is presented.

Coordination Structure in DAI In MAS, coordination structure can be considered as the pattern of interactions between the agents. This pattern can be decided among the agents on the basis of problem-size, complexity, and the number of agents. For example, agents can decide on a simple decentralized structure, multi-hierarchical, or market structure. In this sense, deciding the appropriate structure can be reduced to a problem that has to be solved for a solution domain which consists of different types and levels of coordination structures. Therefore, deciding on a structure as a problem-solving, Ps_s, is a function that maps problem-size (Pz), -complexity (Pc), and the number of agents (N) into a coordination structure (Cs). That is, $Ps_s: Pz \times Pc \times N \rightarrow Cs$.

To determine (or estimate) Pz and Pc assume a problem P that can be decomposed into a set of subproblems (goals) such that $\Pi(P) = G^i = \{g_1, g_2, ..., g_{\lambda_1}\}$, where P is a function that maps a problem, number of agents, models of the other agents by Ag_i, and a set of resources into a set of subproblems. That is, $\Pi: P \times N \times M^1 \times \Re \rightarrow G^1$. Then, each subproblem can be solved by generating a plan. Hence, the problem size is defined as the cardinality of G^i, and its complexity as the cardinality of the plan. The problem-complexity, even though it is related to the solution (plan) of the problem, doesn't require that the solution be known in advance, only how far it is from the initial world state. This can be achieved by developing a problem-complexity estimation function which maps an initial world state and a goal into a number that is an estimate of the problem-complexity.

Coordination Mechanism in DAI Coordination mechanism is represented by a set of heuristics and devices that can be used respectively to reduce and resolve problems associated with each coordination parameter. Heuristics can be domain-independent (such as those listed in Table 8-1), or (and) domain-dependent which can be supplied separately for each application-domain. Some heuristics act as constraints for the decomposition device to produce subproblems that minimize Z. For example, (1) 'avoid generating conflict and common goals'; (2) 'generate goals that can be solved by the smallest self-contained groups of agents'. Solving each subproblem in MAS, as mentioned before, involves three phases: Ps, Sr, and Cr. Hence, heuristics act as constraints that should be satisfied by the three phases. For example, the following act as constraints during the Ps phase: (1) 'select the shortest solution'; (2) 'avoid activities that require resources'; (3) 'avoid interdependent activities'; (4) 'select activities that require less knowledge'. Also, during Sr phase: (1) 'assign problems to the capable agents only'; (2) 'assign problems to the most self-contained group of agents'; (3) 'assign conflict or common problems to the same group of agents'; (4) 'maintain the load-balance condition for resources'. In addition to that some of the heuristics enforce some specifications over M^i such as modeling the other agents capabilities.

To resolve coordination problems associated with each parameter, an agent should be capable of :

1. identifying and reasoning about the type of the parameter; or reasoning about 'which';

2. providing the adequate means or devices for solving problems associated with each parameter; or reasoning about 'how.'

Reasoning about the type of coordination parameters requires mapping these parameters in terms of the MAS formulation. Thus, reasoning about 'which', is simply a function that maps a set of variables $\varphi \subset \text{MAS}$ into others $\gamma \subset \Gamma_p,$. Therefore, for

* $\overline{P_p}$: an agent should be able to reason about its own domain-activities and subproblems; or which $\left(Ac^i, G^i\right) \equiv \overline{P_p}$;

* P_d : an agent should be able to reason about its handicapped activities. To achieve this, an agent can represent its activities in terms of precondition and effect sets, such that $a_j^i = \langle \delta_j^i, \varepsilon_j^i \rangle$; where δ_j^i and ε_j^i are the precondition and the effect sets respectively for the activity; or which $\left(\delta_j^i, \varepsilon_j^i\right) \equiv P_d$;

* P_r : an agent should be able to reason about its set of preconditions that require resources; or which $\left(Ac^i, G^i\right) \equiv P_r$;

* $\overline{P_k}$: an agent should be able to reason about its current problem, precondition set and its local world history; or which $\left(g, \delta_j^i, \omega_T^i | \forall j\right) \equiv \overline{P_k}$;

* P_f, P_c and P_s : an agent should be able to reason about its goals, other agents' goals and solutions; or which $\left(G^i, M_{G_j}^i, M_{Cpl_T^j}^i | \forall j\right) \equiv \{P_f, P_c, P_s\}$.

Reasoning about 'how', a coordination problem can be resolved, can be considered as a function that maps a set of coordination parameters, into appropriate devices. The following is a list of generic devices that are appropriate for their corresponding parameters:

* for $\overline{P_p}$, and P_d, a device called 'Assignment' is required to assign problems to the appropriate agent(s); or how $(\overline{P_p}, P_d) \equiv$ 'Assignment'. Aspects other than reasoning about agent capabilities have to be considered in developing this device. For example, solution efficiency that can be measured by criteria such as the completion time or the load-balance.

* for P_r 'Resources-scheduling' device can be used. This device is somewhat similar to the 'Assignment' device except that some of the resources are consumable; or how $(P_r) \equiv$ 'Resources-scheduling';

* for $\overline{P_k}$ the local knowledge has to be updated either through sensory or communication devices. These devices have to be able to reason about when to update the local knowledge, which part requires updating, and so on; or how $(\overline{P_k}) \equiv$ 'Knowledge-update';

* for P_f a device called 'Conflict-resolution' by which an agent reasons about other agents' goals and solutions are necessary for avoiding the harmful effect of conflicting relationships; or how $(P_f) \equiv$ Conflict- resolution';

* for P_s a device called 'Synchronizer' in which an agent reasons about other agents' goals and solutions in order to coordinate simultaneous goals; or how $(P_s) \equiv$ 'Synchronizer';

* for P_c a 'Redundancy-avoidance' device can be used to reduce the repeated work that will be done by some agents due to common relationship between the agents' goals. One approach is by letting agents reason about other agents' goals and then elect from among themselves agent(s) to carry out the common goals; or how $(P_c) \equiv$ 'Redundancy-avoidance'.

Many different strategies can be used to implement these reasoning capabilities (about 'which' and 'how'). The selection between these strategies is application dependent. One strategy, called **interleaving Ps-Sr-Cr**, considers reasoning about 'which' during Ps phase. For example, agent A in the Assembly scenario can reason about 'which' parameters are encountered (e.g. resource tool-1) during its problem-solving. The Sr phase would then be invoked to reason about 'how' to solve problems associated with these parameters. For instance, get tool-1 which can be achieved by reasoning about the shared resources and the other agents' solutions. Finally, Cr acts as a confirmation phase by transforming coordinated-activities into intentions.

In this strategy Ps proceeds to select the next activity only when the current one is granted by both of Sr and Cr phases, otherwise an alternative solution is suggested. Obviously, interleaved Ps-Sr-Cr guarantees coordination without replanning, however, it is possible that it coordinates activities or solutions that will be rejected locally. For instance, the case in which the local-control of Ps discovers that there is a better solution (a shorter solution) than the one explored. In addition to the new coordination efforts required for the new solution, a re-coordination effort at the society level is required. Also, confirming solutions at the activity level overloads the communication means which makes this strategy unfeasible for many application domains.

An alternative strategy, called **interleaved Ps-Sr**, in which Cr phase is only invoked at the solution level. This strategy avoids the re-coordination effort at the society level, however a replanning is possible. One way of reducing the risk of replanning is to provide Cr with several possible coordinated or partially coordinated plans.

A third strategy, called **plan-level**, tries to minimize the overhead of coordinating activities that are possibly not part of the solution. However, the possibility of replanning is more higher than interleaved Ps-Sr. In this strategy Ps generates partial plan(s) with respect to activities that require coordination. Note, coordination is needed only for interdependent activities. Then Sr transforms these partial plans into corresponding partial coordinated-plans which are in turn transformed to intentions by Cr.

11.5 EXAMPLE: TABLE-ASSEMBLY

In this section the importance of coordination for distributed intelligent systems is shown for a simple manufacturing environment. The example of **Table-assembly** is used to demonstrate the proposed theory of coordination and its implementation approaches in a DAI environment. Apart from being a real-world distributed environment, this 'Table-assembly' scenario satisfies other research objectives.

Firstly, it is complex enough to demonstrate the main issues and problems that are related to the concept of coordination. Secondly, comparisons are easier because it includes features common to other DAI systems designed to solve the coordination problem. For example, by introducing the concept of **reachability** between agents the feasibility of solving problems that are inherently distributed but require coherent solutions can be demonstrated. The concept of coherency is a crucial issue for many DAI systems for which many techniques have been developed such as the organizational structuring and PGP. Thirdly, aspects which are associated with specific existing systems can be addressed in an integrated fashion: the concept of **society** which is the basic concern of ICE; the **interdependency** which is the main focus of contract-based systems; and the **concurrency** concept within the society through the partial-overlapping of agents' capabilities. Finally, the example as a small scale, real distributed environment minimizes the number of assumptions that distorts the real picture of coordination.

11.5.1 Table-Assembly: MAS

A multi-agent system for 'Table-assembly' is defined where the work-cell contains four agents on the workshop-floor, and two on the store-floor. Workshop-agents cannot reach the store-floor, and store-agents cannot reach the workshop-floor, however, all agents can reach the shared-floor. Workshop-agents share common resources such as one drill, two screw-drivers, screws-box, a glue-tube, and the workshop-floor. The store-agents share the store-floor. Given that a table's top-sheet and four legs are available in the workshop-floor, the workshop-agents are asked to assemble a table and deliver it to the shared-floor, then the store-agents deliver the table to the store-floor.

The 'Table-assembly' environment can be easily modeled (or considered) as a MAS which is denoted by $TA = \langle A, O, \Re, \Im \rangle$, where the set of agents is denoted by $A = \{Ag_1, Ag_2, ..., Ag_6\}$, the set of objects by O = { Top-sheet, Leg1, Leg2, Leg3, Leg4, and the set of resources by $\Re$ = { drill, screw-driver1, screw-driver2, screws, screws-box, workshop-floor, shared-floor, store-floor, glue-tube}. For simplicity both computational time and space are not considered as limited resources. Also, it is assumed that all activities have the same processing time. Domain-dependent activities are given in Table 8-2, and the capabilities of each agent in Table 11-3.

Let a problem p = 'Table-assembly', and a decomposition device $\Pi(P) = G = \{G1, G2, ..., G6\}$, where Gi is a sub-goal to be assigned to Ag_i. The details of each of these sub-goals are given in Table 11-4. It can be shown that G1, G2, and G3 are independent goals in the sense that any possible world that satisfies G1 can be changed to any possible world that satisfies G2 and G3 without disturbing G1. Therefore G1 || G2 as well as G1 || G3, and the same can be applied for G2 and G3 for G2 || G3. Given that a table-leg can only be affixed on the workshop-floor. This makes G1, G2, and G3 intermediate goals. Accordingly, starting with any possible world that satisfies G4 it is impossible to change this world to another world that satisfies G1; G2; and G3 without disturbing G4, and hence G4 <> g; g ∈ { G1; G2; G3} . Also, under the assumption that an entity cannot occupy two different special locations at the same time, which makes G4 an intermediate goal. Since starting with any possible world that satisfies G4 it is impossible to change it to another world that satisfies G5 without disturbing G4, therefore G4 <> G5. Similarly it can be shown that G5 <> g; g ∈ {G1; G2; G3}.

At the initial state, $t = 0$, obviously $Pl_o^i = I_o^i = \phi$, and the local world history is ω_o^i = {'Leg1 at workshop-floor', 'Leg2 at workshop-floor', 'Leg3 at workshop-floor', 'Leg4 at workshop-floor', 'Top-sheet at workshop-floor', 'store-shared-door is closed'}. Also let $M^i = \{\langle M_{G^l}^i, M_{Cpl_T^l}^i, M_{Ac^l}^i \rangle | 1 \le l \le 6, i \ne l\}$. For example, $M_{G^2}^1$ = { 'Leg2 fixed at Loc2'} is the model of agent Ag_2's goals by agent Ag_1, with $M_{Cpl_o^2}^1 = \phi$ at $t = 0$, and $M_{Ac^2}^1$ = {tight-screw, fix-by-screw, move-table}.

Using the above example the feasibility of the proposed theory is demonstrated throughout the following subsections. In the first, the reasoning capabilities about 'which'

Table 11-2: Domain-dependent activities.

Activity	precondition set	effect set
drilling	{'got a drill'}	{'hole at loc'}
tight-screw	{'hole at loc', 'got a screw', 'got a screw-driver'}	{'screw is tight at loc'}
fix-by-screw	{'entity at loc', 'screw is tight at loc'}	{'entity fixed at loc'}
fix-by-glue	{'entity at loc', 'glue at loc'}	{'entity fixed at loc'}
apply-glue	{'got a glue-tube'}	{'glue at loc'}
bring	{'entity at i-loc'}	{'entity at d-loc'}
open-door	{'door-name is closed'}	{'door-name is open'}
move-table	{'leg1 fixed at loc1', 'leg2 fixed at loc2', 'leg3 fixed at loc3', 'leg4 fixed at loc4', 'door-name is open'}	{'Table at name-floor'}

Table 11-3. Agents' domain-dependent capabilities.

Agent name	Domain-dependent Capabilities (Act_i)
Ag_1	{apply-glue, bring, drilling, fix-by-screw, fix-by-glue}
Ag_2	{tight-screw, bring, fix-by-screw, move-table}
Ag_3	{tight-screw, bring, drilling, fix-by-screw}
Ag_4	{open-door, move-table}
Ag_5	{open-door, move-table}
Ag_6	{open-door, move-table}

Table 11-4. Sub-goals for $\mathcal{P}$.

Ag_i's sub-goal	Sub-goal details
G 1	{'Leg1 fixed at Loc1'}
G 2	{'Leg2 fixed at Loc2'}
G 3	{'Leg3 fixed at Loc3', 'Leg4 fixed at Loc4'}
G 4	{'Table at shared-floor'}
G 5	{'Table at store-floor'}
G 6	{'store-shared-door is open'}

Coordination of Distributed Intelligent Systems 253

and 'how' are demonstrated within the phases of problem-solving. In the second, the proposed strategies for implementing these reasoning capabilities are demonstrated.

11.5.2 Coordination: Phases of Problem Solving

For simplicity and without loss of generality, the main issues of the coordination problem that have been discussed throughout the chapter are demonstrated for achieving G1 and G5 by Ag_1 and Ag_5 respectively.

Assume that a problem can be solved in a number of 'ways' which are possible solutions not necessarily generated by Ps and perhaps equivalent to a 'plan' context.

Consider achieving G1 for Ag_1:

* 'way1' = { a_{11}^1 =' get a drill', a_{12}^1 ='drilling at Loc1', a_{13}^1 ='bring Leg1 to Loc1', a_{14}^1 = 'access screws-box', a_{15}^1 ='get a screw', a_{16}^1 = 'get screw-driver', a_{17}^1 ='tight-screw at Loc1', a_{18}^1 ='fix-by-screw Leg1 at Loc1'}; or

* 'way2' = { a_{21}^1 ='get glue-tube' a_{22}^1 ='apply glue at Loc1', a_{23}^1 ='bring Leg1 to Loc1', a_{24}^1 ='fix-by-glue Leg1 at Loc1'}.

Further, assume that the planning mechanism is plan-based, and 'way1' as well as 'way2' can be used to represent a possible solution domain for G1 as shown in Figure 11.3. Box-1 contains activities that can be delegated either to Ag_2 (represented by branching through node a_{17}^1), or to Ag_3 through node $a_{17}^{'1}$.

Assume that Ps of Ag_1 starts to explore P1 in Figure 11.3. Using the capability of reasoning about 'which', it can be determined that a_{18}^1 is a dependent activity. If the precondition 'screw is tight at Loc1' does not hold, then it is impossible for Ag_1 to assert it. In this case a dependent-activity denoted by a_{17}^1?='screw is tight at Loc1?' is issued. The same for a_{12}^1 where a resource-activity a_{11}^1? ='got a drill?' is also issued. Then, Sr explores 'how' to resolve problems associated with these activities.

For a_{17}^1? the 'Assignment' device is invoked to reason about $M^1_{Ac^l}$. In this case there will be two candidates Ag_2 and Ag_3. Using the load-balance criterion and the self-contained heuristic (both Ag_1 and Ag_2 as well as Ag_3 and Ag_4 can be categorized as self-contained groups) Ag_2 can be elected. In general, Sr would generate either a coordinated-plan (a_{17}^1? : Ag_2) where a_{17}^1? is delegated to Ag_2 or partial coordinated-plan as (a_{17}^1? : Ag_2; Ag_3) where a_{17}^1? can be delegated either to Ag_2 or Ag_3. Notationally, the agents list represents the preference order for the servant agent.

To resolve a_{11}^1? Sr invokes 'Resource-scheduling' to reason about the shared-resources and to keep track of the resources schedule. For instance, assume that the status of the drill at time t is free. However, by the reasoning about $M^1_{CpI_T^3}$ it can be concluded that it is possible that the drill will be used by Ag_3 at time t. In this case, a partial coordinated-plan can be generated such as $(a^1_{11}?: \{ \text{Drill; t; } Ag_3 \}; \{ \text{Drill; t + 1; } \phi \})$ where the drill can be requested at time t at which it is possible to be requested by Ag_3, or at time $t + 1$ at which it is expected to be free.

The third phase Cr transforms the coordinated-plans to intentions. For instance, $(a^1_{17}?: Ag_2)$ has to be confirmed by negotiating with Ag_2 to guarantee correctness and coherency. This determines whether Ag_2 is willing to accept it; if it accepts then when; and so on. Similarly if $(a^1_{11}?: \{ \text{Drill; t + 1; } \phi \})$ is generated then Cr has to confirm the new schedule of the drill to Ag_1 and to the other agents as well.

Nevertheless, if Ps has considered the 'avoid dependent activities' heuristic in the first place then P2 in Figure 11.3 would be selected over P1 because it requires less coordination effort. The only coordination activity for P2 is $a^1_{21}? = $ 'got a glue.'

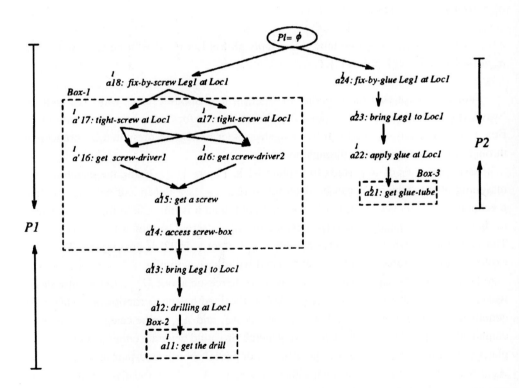

Figure 11.3. A solution domain for G1.

Coordination of Distributed Intelligent Systems

To demonstrate the reasoning capabilities about the other parameters consider achieving G5 = 'Table at store-floor' by Ag_5, given the possible plan space shown in Figure 11.4 where

* a_1^5 = {'open-door store-shared-door'},

* a_2^5 = {'move-table from shared-floor to store-floor'}.

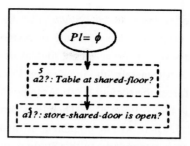

Figure 11.4. A solution for G5.

Assume that Ag_5 does not know the truth value of 'Table at shared-floor', because this proposition is not present in Ag_5's memory (not available knowledge). In this case, Ag_5's Ps generates a knowledge-activity a_2^5? = 'Table at shared-floor?' for which Sr invokes 'Knowledge-update' to resolve Ag_5? One way of achieving this is by broadcasting an inquiry through Cr.

Since Sr concludes that G5 $\diamond$ g; g $\in$ {G1; G2; G3; G4}, then it invokes 'Conflict-resolution' to reason about $M_{G^l}^5$ and $M_{Cpl_T^l}^5$; $1 \leq l \leq 4$. A temporal conflict will be determined because G5 cannot be reached before or at the same time as G1; G2; G3, and G4. Therefore, some sort of inter-relationship is enforced between the activities of different plans such as 'fix-by-screw Leg1 at Loc1' $\Rightarrow$ 'move-table from workshop-floor to shared-floor' $\Rightarrow$ 'move-table from shared-floor to store-floor' which has to be confirmed with the involved agents through Cr.

A common-activity also will be discovered by Ag_5 because a_1^5? = 'store-shared door is open?' is equivalent to G6. In turn 'Redundancy-avoidance' is invoked to determine who will carry out this activity. In this case Ag_6 will be elected and to be confirmed with Ag_5 through Cr. Therefore, a_1^5 will no longer be part of Cpl_T^5.

11.5.3 Coordination: Strategies

In this section the feasibility and the main differences between interleaved Ps-Sr-Cr, interleaved Ps-Sr, and plan-level strategies are demonstrated. Using solution domain shown in Figure 11.3 for achieving G1 by Ag_1. It is assumed that Ag_1's Ps uses some heuristic which prefers P1 over P2 to be explored. First consider when the interleaved Ps-Sr-Cr strategy is applied. Also, assume that Ps of Ag_1 generates a_{17}^1? for which Sr generates (a_{17}^1? : Ag_2). In turn Cr has to confirm it, and so on for each generated activity. For simplicity assume that there is no back tracking encountered until a_{12}^1 has been reached. If at this point the heuristic of Ps discovers that P2 is cheaper than P1 then both new coordination cycles for P2 and a society re-coordination phase has to be reconsidered. This can be a very costly process because resolving a^1_{17}? by Ag_2 requires generating plan steps from a^1_{17} to a^1_{14} which in turn affect the schedule of resources such as screw-drivers and screws-box, and the activities of other agents such as of Ag_3.

Alternately, the interleaved Ps-Sr strategy can be used to avoid this re-coordination effort at the society level. Assume that Sr generates partial coordinated-plans (a^1_{17}? : Ag_2; Ag_3) for solving a^1_{17}?. Then Ps is permitted to generate the next activity as shown in Figure 11.5. There will be no plan confirmation until a coordinated-plan or a partial coordinated-plan is generated. In this strategy no inter-commitments are made until a plan is locally chosen. However, the possibility of unnecessary coordination still holds but not at the society level.

So, alternatively, plan-level strategy can be used to reduce the unnecessary efforts of Sr. Assume that the Ps generates a partial-plan for G1, shown in Figure 11.6a, for which Sr elaborates as shown in Figure 11.6b, then, Cr will confirm the complete plan as shown in Figure 11.6c using the same procedures that are explained above. However, the possibility of replanning is higher in the plan-level strategy than in the interleaved Ps-Sr strategy.

The example shows how a real-world environment can be represented as a DAI system. The feasibility of the proposed 'Theory of Coordination' and its implementation approaches for modeling multi-agent systems also have been demonstrated using the 'Table-assembly' scenario.

11.6 CONCLUSION

Coordination is an interdisciplinary concept, and moreover it is central for distributed intelligent systems. Clearly, real-world problems cannot be solved in distributed systems without coordination. A survey of the research literature for coordination with special focus on DAI leads to the conclusion that: 'coordination' and many related concepts are ill-defined, and there is no complete body of coordination theory that can be used to analyze when, which and why a coordination technique is appropriate. This has been the main focus of this research.

In this work the concept of coordination has been determined in terms of parameters and components. Parameters are informally types of interdependencies and uncertainty. Components are 'structure' and 'mechanism'. Coordination structure is either centralized or decentralized, however, it is easy to build on these two basic types to fit problem characteristics of size and complexity with respect to its bounded-capabilities. Coordination mechanism is comprised of a collection of devices to handle different types and topologies of interdependencies, and uncertainty of task-knowledge and control. Uncertainty in coordinated-control, by definition, is the most crucial for coordination. This chapter has introduced a theory of coordination for which a quantitative representation of

Coordination of Distributed Intelligent Systems

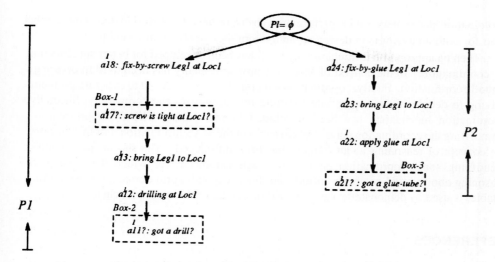

Figure 11.5. A solution domain for G1 by interleaved Ps-Sr.

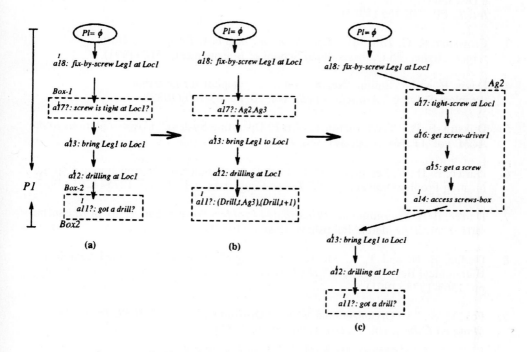

Figure 11.6. A solution for G1 considered by plan-level strategy.

coordination parameters and a performance measure have been formulated. This can be used for both analyzing and designing an appropriate coordination techniques.

Then an appropriate formulation for MAS has been developed for better specification of coordination. In the context of MAS, the proposed theory has been used to decide on a **good** coordination. Firstly, coordination structure has been reduced to structure problem-solving to decide on an appropriate structure that minimizes the complexity. Secondly, coordination mechanism has been developed in two parts. The first part requires formulating the coordination parameters in terms of the MAS formulation. For this purpose, the concept of reasoning about 'which' has been introduced. The second part requires identifying some coordination devices for each parameter, for which the concept of reasoning about 'how' has been introduced. Finally a real-world domain, for assembling a table, is used to demonstrate the proposed theory and solution approaches.

REFERENCES

1. Bond, A. H. and L. Gasser. *Readings in Distributed Artificial Intelligence.* Morgan Kaufmann, San Mateo, Ca. (1988).

2. Corkill, D. D. and V. R. Lesser. "The Use of Meta-Level Control for Coordination in a Distributed Problem Solving Network." *Prec. of the 8th Inter. Joint Conf. on Artif. Intell.*, PP. 748-755 (1983).

3. Crowston, K. G. *Towards a Coordination Cookbook: Recipes for Multi-Agent Action.* Ph.D. thesis Sloan School of Management, MIT (1991).

4. Davis, R. and R. Smith. "Negotiation as a Metaphor for Distributed Problem Solving." *Artificial Intell.*, vol. 20, pp. 63-109 (1983).

5. Dijkstra, E. W. "The Structure of T.H.E. Operating System." *Communications of the ACM.* vol. 11, No. 5, pp. 341-346, 1968.

6. Durfee, E. H., V. Lesser, and D. D. Corkill. "Coherent Cooperation among Communicating Problem Solvers." *IEEE Trans. Computers.* vol. 36, pp. 1275- 1291 (1987).

7. Durfee, E. H. *Coordination of Distributed Problem Solvers.* University of Massachusetts, Kluwer Academic Publishers, Boston (1988).

8. Durfee, E. H. and T. A. Montgomery. "Coordination as Distributed Search in a Hierarchical Behavior Space." *IEEE Trans. on Syst., Man, and Cyber.,* vol. 21, no. 6. pp. 1363-1378 (1991).

9. Fox, M. S., "An Organizational View of Distributed Systems," *IEEE Trans. on Syst., Man and Cyber.*, vol. 11, No. 1, pp. 70-79 (1981).

10. Galbraith, J. R., Organization Design, Addison-Wesley (1977).

11. Gasser, L. and M. Huhns. (Eds.) *Distributed Artificial Intelligence.* vol. II, Morgan Kaufmann, San Mateo, California(1989), pp. 259-290.

12. Gasser, L., N. F. Rouquette, and R. W. Hill. " Representing and Using Organizational Knowledge in DAI systems." *Distributed Artif. Intell.* vol. II, .L Gasser and M. Huhns, ed. (1989), pp. 55-78.

13. Gasser, L. "Social Conceptions of Knowledge and Action: DAI Foundations and Open Systems Semantics." *Artificial Intelligence.* Vol. 47 (1991), pp. 107-138.

14. Georgeff, M. "Communication and Interaction in Multi-Agent Planning," *Proc. on AAAI (1983)*, pp. 125-129.

15. Georgeff, M. "Many Agents are Better than One." *The Frame Problem in Artif. Intell.*, F. Brown, ed., Morgan Kaufmann (1987), pp. 59-75.

16. Hewitt, C. "Offices are Open Systems." *ACM Trans. Office Info. Sys.* vol. 4, No. 3. (1986), pp. 271-287.

17. Hewitt, C. "Open Information Systems Semantics for Distributed Artificial Intelligence." *Artificial Intelligence.* Vol. 47 (1991), pp. 79-106.

18. Kamel, M. and A. Syed. "An Object-Oriented Multiple Agent Planning System." *Distributed Artificial Intelligence.* vol. II, L. Gasser and M. Huhns (Eds.), Morgan Kaufmann, San Mateo, California (1989), pp. 259-290.

19. Kamel, M. and H. Ghenniwa. "On Solving the Assignment Problem for Partially Overlapped Systems." *Proc. International Symposium on Robotics And Manufacturing* (1992).

20. Kamel, M. and H. Ghenniwa. "Partially Overlapped Multi-Machine Systems: The Assignment Problem." Submitted to *IEEE Trans. on Systems, Man, and Cybernetics.*

21. Kornfeld, W. A. and C. E. Hewitt. "The Scientific Community Metaphor." *IEEE Trans. on Systems, Man, and Cybernetics.* Vol. SMC-11, No. 1 (1981), pp. 25-33.

22. Lawler, E., J. Lenstra, A. Rinnooy Kan, and D. Shmoys. " Sequencing and Scheduling: Algorithms and Complexity." *Tech. Report BS-R8909, Center for Mathematics and Computer Science*, Amsterdam, The Netherlands (June 1989).

23. Lesser, V. R. and D. D. Corkill. "Functionally Accurate, Cooperative Distributed Systems." *IEEE Trans. on Systems, Man, and Cybernetics.* Vol. SMC-11, No. 1 (1981), pp. 81-96.

24. Malone, T. W. " What is Coordination?" *Working Paper #2051-88.* MIT, Salon School of Management (1988).

25. Malone, T. W. and K. Crowston. "What is Coordination Theory and How can it Help Design Cooperative Work Systems?" *Proceedings of the third Conf. on Computer Supported Cooperative Work*, in D. Tatar, ed. (1990), pp. 357-370, 1990.

26. Malone, T. W. and K. Crowston. "Toward an Interdisciplinary Theory of Coordination." *Thec. report CCS TR# 120*. Salon School of Management, MIT (1992).

27. March, J. G. and H. A. Simon. *Organizations*. John Wiley and Sons, New York (1958).

28. Rosenchein, J. and M. Genesereth. "Deals among Rational Deals." *Proc. of the Int. Joint Conf. of Artif. Intell.* (1985), pp. 91-99.

29. Rosenchein, J. "Rational Interaction: Cooperation among Intelligent Agents." *Ph.D. Thesis*. Stanford Univ. (1986).

30. Rosenchein, J. and J. Breese. "Communication-Free Interactions among Rational Agents: A Probabilistic Approach." *Distributed Artificial Intelligence*. L. Gasser and M. Huhns, eds., Morgan Kaufmann, San Mateo, California (1989) pp. 99-118.

31. Simon, H. A. *Models of Man*. Part IV in Wiley, New York, pp. 196-279, 1957.

32. Sycara, K., "Multi-Agent Compromise via Negotiation." *Distributed Artif. Intell.* Vol. 2, Les Gasser and M. Huhns, ed. (1989), pp. 119-138.

33. Tanenbaum, A. S. *Computer Networks*. Prentice-Hall, Englewood Cliffs, NJ (1981).

34. Thompson, J. D. *Organizations in Action*. McGraw-Hill, New York (1967).

35. Williamson, O. E. *Markets and Hierarchies: Analysis and Antitrust Implications*. The Free Press Publishers, New York (1975).

36. Zlotkin, G. and J. Rosenchein. "Incomplete Information and Deception in Multi-Agent Negotiation." *Proc. of the 12th Inter. Joint Conf. on Artif. Intell.* (1991), pp. 225-231.

12

COLLABORATIVE WORK BASED ON MULTIAGENT ARCHITECTURES: A METHODOLOGICAL PERSPECTIVE

B. Moulin, Université Laval, Québec, CANADA

L. Cloutier, Division du Commandement et Contrôle Centre de Recherche Pour la Défense, Québec, CANADA

Team work is one of the most common forms of work organization in human societies, from the most primitive through to the most industrialized. In this chapter, we begin by investigating how several techniques borrowed from distributed artificial intelligence can be used to build multiagent systems to support collaborative work. Then, we describe the Multi-Agent Scenario-Based method, a multiagent system design approach based on the analysis and design of scenarios involving human and artificial agents. This method is well suited to model systems supporting human computer cooperation work. We are particularly interested in group interactions involving persons and machines where artificial agents are autonomous: they are able to accomplish various activities by themselves and manage their own local data bases. In certain situations an agent may require the help of other agents, consequently initiating group activities. Agents do not only interact with one another, but they also have access to common data representing the world in which they evolve. These data are managed by an agent called the object server which gathers active objects. Different design techniques are proposed to describe scenarios and agents' behaviour: behaviour diagrams, data models, transition diagrams, object life cycles and object behaviour diagrams.

12.1 INTRODUCTION

In human organizations many activities are done by groups of persons usually acting cooperatively. When these activities correspond to data exchanges, we can develop classical information systems which enable users to share data through data bases and transfer information through communication networks. When group activities involve decision making, planning and knowledge sharing, classical information systems are no longer adequate and people currently rely on face-to-face meetings. During the last decade, in order to assist some of their users in making decisions, organizations have introduced several expert systems which are able to solve problems on the basis of some explicit knowledge (usually rules) stored in their knowledge bases. Most expert systems have been designed to support users on an individual basis. However, in several cases companies require systems which support activities involving the collaboration of a group of persons and knowledge-based systems (Chang 1987).

Team work is one of the most common forms of work organization in human societies, from the most primitive through to the most industrialized. Team work implies several people associating themselves in order to reach a common goal, each person playing a given role in the team according to her skills and abilities. Team workers must coordinate their actions in order to perform their tasks efficiently, to avoid duplication of efforts and potential conflicts. There are several ways to coordinate group activities: in a "centrally-driven team", one person, playing a manager's role, directs and controls all team members' activities; in a "consensus-driven team", team members can negotiate together in order to make decisions and elaborate plans that will be distributed among them. Coordination may also arise from the characteristics of the artefact (an object, a building, a machine, a document, etc.) that team members intend to build together: in that case the artefact blueprints help team members determine who will undertake each necessary activity and when this will be done.

Computers are used to support various kinds of work activities such as word processing, spreadsheet calculations, data management, design and communication. They provide users with powerful tools to perform their individual tasks, but team coordination usually still relies on person-to-person interactions. However, there is a growing interest in computer systems which support group activities. This recent research field has been coined "*Computer-Supported Cooperative Work*" (CSCW) (Greif 1988) (ACM 1991). Hence, our perspective of team work can be enlarged: a team will be considered as a group of persons and computers that collaborate together in order to solve a common problem or to reach a common goal. Each team member will be called an *agent* that can be either human or artificial. Artificial agents are linked together by a telecommunication network and interact with human agents through communication interfaces. In such a work organization, tasks are distributed among artificial and human agents in order to take advantage of the abilities of each team member. Artificial agents may thus be considered as intelligent assistants of their users (Chang 1987): they relieve human agents from tedious and time-consuming group activities such as searching and obtaining information from other team members, synchronizing agendas to find a time slot for a meeting and scheduling activities for creating a new artefact for a customer.

Artificial and human agents working together to achieve a common goal form a *multiagent system* that can be designed using techniques developped in Distributed Artificial Intelligence (DAI). Generally, DAI investigates the possibility of solving difficult problems by distributing parts of the problem among several intelligent agents who cooperate or negotiate in order to solve the problem at hand (Durfee et al. 1989, Chaib-draa et al. 1992, Gasser et al. 1989). The most common way of approaching DAI problems is to focus on the planning and coordination issues: agents are considered as autonomous intelligent planners and an environment is designed in order to enable them to exchange knowledge (intermediate goals or plans, beliefs, facts etc.) to incrementally reach a solution. Agents cooperate in order to build an acceptable plan or negotiate in order to reach an agreement that will provide a solution.

In order to illustrate how human and artificial agents may collaborate as a team, we will take the simple example of a fictitious workshop where craftsmen create guitars of various kinds *(acoustic, electric)* and of various qualities (medium, good, professional). There are four men working in this workshop: an administrator who processes customer orders, launches internal fabrication orders and is responsible for delivering guitars to customers and receiving their payments; an assembler who assembles the main parts of the guitar (body, neck, strings); an electronics specialist who incorporates electrical systems into electric guitars, fine-tunes them and tests each guitar; and a fine-work specialist who applies finishing touches to the guitars (paintings, decorations). Since this is a famous workshop, orders are numerous and each worker is very busy, skillfully performing his own tasks. Coordination tasks (such as scheduling production orders, passing fabrication instructions to craftsmen, organizing meetings and following each order along the production line) are time-consuming and considered by craftsmen as a loss of their precious time. In order to relieve craftsmen from administrative tasks, each worker is given an artificial assistant (on a microcomputer) which will be responsible for performing coordination tasks in relation with other craftsmen artificial assistants. In the following sections, we will investigate how several DAI techniques can be used to build such a multiagent system for supporting collaborative work.

12.2 DISTRIBUTED ARTIFICIAL INTELLIGENCE

12.2.1 Introduction

DAI research has been expanding quickly since the early eighties. It is not our intent here to present a detailed review of this vast research field. Instead, we will focus on some issues that should be considered when building a multiagent system.

A *multiagent system (MAS)* can be defined as "a loosely-coupled network of problem solvers that work together to solve problems that are beyond their individual capabilities" (Durfee et al. 1989). These problem solvers, often called *agents*, are autonomous, potentially preexisting the MAS and may be heterogeneous (characterized by various degrees of problem solving capabilities).

Durfee et al. (1989) observe that many applications are inherently distributed. Some applications are spatially distributed (such as interpreting and integrating data obtained from spatially distributed sensors, or controlling a group of robots). Other applications are functionally distributed (such as a group of experts with different specialities collaborating to solve a difficult problem).

A MAS has significant advantages over a single, monolithic, centralized problem solver: *faster problem solving* by exploiting parallelism; *decreased communication* by transmitting only high-level partial solutions to other agents rather than raw data to a central site; *more flexibility* by having agents with different abilities dynamically team up to solve current problems; and *increased reliability* by allowing agents to take on responsabilities of agents that fail.

Here are some application domains where DAI techniques are applied:

- *Distributed interpretation* applications collect, interpret and integrate distributed data to elaborate a semantic data model, as in the DVMT project (Lesser Corkhill 1983).

- *Distributed planning and control* applications involve developing and coordinating the actions of agents to perform desired tasks. Application domains include air-traffic control (Cammarata et al. 1983), distributed process control in manufacturing (Parunak 1987) and resource allocation/control in a long-haul communication network (Adler et al. 1989).

- *Cooperating expert systems* applications allow several expert systems to work together to solve a common problem. A typical example is the cooperation between expert systems in engineering (Bond 1989).

- *Computer-supported cooperative work* applications in which agents can assist persons in managing distributed data, filtering information and coordinating activities among groups of users (Chang 1987, Huhns et al. 1987, Malone 1990, Pan and Tenenbaum 1991). Artificial agents, behaving as human personal assistants, improve team coordination by solving coordination problems such as scheduling meetings or routing messages to relevant people. Besides relieving users from several coordination tasks, artificial agents can work in parallel and "behind the scenes" (in the background or at night) to share information and improve coordinated decisions.

Several main issues should be considered when we study or design a multiagent system: the characteristics of individual agents; the characteristics of the group of agents; planning issues for the group of agents and communication issues. These points are examined in the following sections.

12.2.2 Characteristics of Individual Agents

We distinguish two main agent categories: human agents and artificial agents. In order to behave autonomously, agents should ideally possess several abilities: perception and interpretation of incoming data and messages, reasoning based upon their beliefs, decision making (goal selection, solving goal interactions, reasoning on intentions), planning (selection or construction of action plans, conflict resolution, resource allocation), and the ability to execute plan including message passing. However, in practical multiagent systems, agents may be characterized by various degrees of problem solving capabilities. We distinguish reactive, intentional and social agents.

A *reactive* agent reacts to changes in her environment or to messages from other agents. She is not able to reason about her intentions (goal manipulation). Her actions are performed as the result of triggering rules or of executing stereotyped plans: updating the agent's fact base (or belief space), and sending messages to other agents or to the environment. First-generation expert systems (composed of a knowledge base containing rule sets, a fact base and an inference engine) are typical examples of reactive agents. Within a MAS, they should also be able to communicate with other agents: choosing and sending, or receiving and interpreting relevant messages according to the current situation.

An *intentional* agent is able to reason on her intentions and beliefs, to create plans of actions and to execute them. But she does not possess explicit models of other agents: she is not able to reason on other agents' beliefs, intentions and plans. Intentional agents can be considered as planning systems (Wilensky 1983, Wilkins 1988, Allen et al. 1991, Lizotte et Moulin 1990, Von Martial 1992): they can select their goals (according to their motivations) and reason on them (detection and resolution of goal conflicts and coincidences), select or create plans (action scheduling), detect conflicts between plans (resource allocation for instance) and execute and revise plans, if necessary. In a MAS, intentional agents coordinate together by exchanging information about their beliefs, goals or actions; this information being eventually incorporated in their plans

In addition to intentional agent capabilities, a *social* agent possesses explicit models of other agents. Hence, she must be able to maintain these models (updating beliefs, goals and eventually plans), to reason on the knowledge incorporated in these models (intentions, commitments, expectations, anticipated reactions and hypothetical behaviours), to make her decisions and create her plans with respect to other agents' models. The level of complexity of the models a social agent has of other agents (in the organization) is relative to the sophistication of that social agent.

12.2.3 Characteristics of the Group of Agents

Several multiagent system characteristics are directly related to the group of agents: group organization, cooperation, negotiation and agent roles. Planning and communication will be discussed in sections 12.2.4 and 12.2.5.

Malone (1990) proposes a comprehensive study of *group organization*. "A group of agents is an organization if they are connected in some way (arranged systematically) and their combined activities result in something better (more harmonious) than if they were not connected. An organization consists of: a group of agents; a set of activities performed by agents; a set of connections among agents; and a set of goals or evaluation criteria by which the combined activities of the agents are evaluated". Hence, group organization depends upon the capacity of agents to coordinate their activities.

Malone (1990) notes that two of the most fundamental components of *coordination* are the *allocation of scarce resources* and the *communication of intermediate results*. "For example, *synchronizing interdependent activities* involves both of these components. If one activity requires as inputs the results of other activities, then synchronizing the communication of intermediate results is required. If nondivisible scarce resources (such as assembly line time) must be shared, then allocating these resources requires spreading demands out over time". . . "In order to fulfill the allocation and communication functions of coordination, agents must be connected to each other. We will primarily be concerned here with information connection rather than physical connections. In order to communicate intermediate results, information must be transferred over these *information links* . . . In order to allocate shared resources, activities must be able to transfer control over the shared resources. This ability often leads to the ability to prescribe behaviour, as well. Therefore, it is sometimes useful to distinguish *control links* as a special kind of information link that carries information the recipients are motivated to follow as instructions."

Mintzberg (1979) considered three fundamental coordination processes. *Mutual adjustment* is the simplest form of coordination. It occurs whenever two or more agents agree to share resources to achieve some common goal. Agents usually must exchange a lot of information and make many adjustments in their own behaviour, depending on the behaviour of other agents. In coordination by mutual adjustment, no agent has any prior control over the others, and decision making is a joint process. Coordination in peer groups and in markets is usually a form of mutual adjustment.

Direct supervision occurs when two or more agents have already established a relationship in which one agent has some control over the others. This prior relationship is usually established by mutual adjustment (as when an employee or subcontractor agrees to follow directions given by a supervisor). In this form of coordination, the supervisor controls the use of shared resources (such as human labour, computer processing time and money) by the subordinates and may also prescribe certains aspects of their behaviour.

In some cases the supervisor coordinates by *standardization*, establishing standard procedures for subordinates to follow in a number of situations. Routine procedures in companies or computer programs are examples of coordination by standardization.

Malone (1990) suggests that using these fundamental coordination processes, it is possible to construct sophisticated systems for coordination, two of the most prevalent ones being hierarchies and markets. *Hierarchies* are based on the process of direct supervision. Mutual adjustment works well in small groups. However, as the size of the groups (and the number of tasks) grows, the number of information links and the amount of information to be exchanged become prohibitive very quickly. A large group can be efficiently divided into subgroups if most of the necessary information transfer can occur within subgroups, and if the few interactions between subgroups can be handled by supervisors. Hence, a hierarchy may be implemented within the group. Subgroups may be coordinated either by mutual adjustment or by hierarchical control, depending on the application domain and the task characteristics. However, it seems that most human hierarchical organizations are unable to function practically without "informal" nonhierarchical communication. These group organizations are called *augmented hierarchies*. Hierarchies can be augmented by standardized communication procedures linking different subgroups, by shared data repositories or by explicitly setting lateral relationships between sub-groups.

Markets can be considered as another form of group organization based on mutual adjustment. Agents, each of whom controls scarce resources (such as labour, raw material, goods and money), agree to share some of their respective resources to achieve some mutual goal. Valued resources are exchanged, either with or without explicit prices. Once a contract has been made, there is an agreement in which the buyer becomes the supervisor of the supplier.

Many organizations implement mixed coordination processes partly based on mutual adjustment, direct supervision and standardization. These organizational frameworks are used as a source of inspiration for choosing adequate organizational structures when developing multiagent systems.

Implementing efficient ways of *cooperation* among agents is a central issue for MAS development. Durfee et al. (1989) propose four generic goals for cooperation among a group of agents: increase the task completion rate through parallelism, increase the set or scope of achievable tasks by sharing resources (information, expertise, physical devices, etc.); increase the likelihood of completing tasks by undertaking duplicate tasks, possibly with different methods of performing those tasks and decrease the interference between tasks by avoiding harmful interactions. Network control is difficult in a network of agents because limited inter-agent communication restricts each agent's view of network problem-solving activity. Durfee et al. (1989) suggest that effective MAS network control involves balancing efficient use of communication and processing resources, high reliability, responsiveness to unanticipated situations and solution quality based on application-specific criteria.

Negotiation plays a fundamental role in human cooperative activities, allowing people to resolve conflicts that could interfere with cooperative behaviour. Durfee et al. (1989) define negotiation as the process of improving agreement (reducing inconsistency and uncertainty) on common viewpoints or plans through the structured exchange of relevant information. By analogy with negotiation processes in human organizations, several negotiation protocols have been proposed for MASs.

The *Contract-Net Protocol* (Smith Davis 1981) has been one of the most influential negotiation approaches proposed for MASs. It was inspired by contracting processes in human organizations. Agents coordinate their activities through contracts to accomplish specific goals. Contracts are elaborated in a top-down manner. An agent, acting as a *manager*, decomposes its contract (the task or problem it was assigned with) into sub-contracts to be accomplished by other *potential contractor* agents. For each sub-contract the manager announces a task to the agent network. Agents receive and evaluate the announcement. Agents with appropriate resources, expertise and information reply to the manager with bids that indicate their ability to achieve the announced task. The manager evaluates the bids she has received and awards the task to the most suitable agent, called the contractor. Finally, manager and contractor exchange information together during the accomplishment of the task. A contractor may also decompose its sub-contract into other sub-contracts and becomes a manager for its subcontract. Smith and Davis investigated the performance of the Contract-Net Protocol on different applications, as for example distributed interpretation where a network should track vehicles over a large geographical area. Parunak (1987) used this protocol to develop a manufacturing control system.

In another influential negotiation protocol, Cammarata et al. (1987) studied *cooperation strategies for resolving conflicts* among plans of a group of agents. They applied these strategies to the air-traffic control domain, where the goal is to enable each agent (aircraft) to construct a flight plan that will maintain a safe distance with each of the aircrafts located in its flying area and satisfy other constraints such as reaching the desired destination with minimum fuel consumption. These authors choosed a policy they called *task centralization*, in which agents involved in a potential conflictual situation (airplanes becoming too close according to their current headings) choose one of the agents involved in the conflict to resolve it. The agents act as a centralized planner (see Section 12.2.5) to develop a multiagent plan that specifies the concurrent actions of all the planes involved. The agents use negotiation to decide which of them is most able to do the planning. This ability can be evaluated on the basis of different criteria permitting to identify for instance the most-informed agent or the most-constrained agent.

The preceding negotiation protocols assumed that agents are cooperative, hence pursuing some common goal. Sycara (1989) developed a system that resolves adversarial conflicts in the domain of labor relations. Three agents are involved: the employer, the employees' union representative and a mediator. Employer and union agents usually have conflicting goals. The mediator generates offers and counter-offers in order to narrow the differences between the parties.

As shown in the preceding discussion, agents may play various roles in a multiagent system. When designing such a system, it is important to identify the roles that can be assigned to agents, as well as how and when these roles are assigned.

12.2.4 Interaction and Communication Between Agents

Agents may interact together either through explicit linguistic actions (communication) or through non-linguistic actions modifying the world in which they are acting. Communication enables agents to exchange information and to coordinate their activities. In multiagent systems two main strategies may be used to support communication. Agents can exchange *messages* directly or they can access a shared data repository in which information can be posted and retrieved. This data repository may be structured into several layers and is usually implemented using a *blackboard* (Nii 1986). These two communication modes may be combined in complex systems. Each agent is composed of several subsystems (or "sub-agents") which exchange information using a local blackboard; agents communicate together by exchanging messages. Several MASs adopted this kind of architecture. The Distributed Vehicule Monitoring Testbed (DVMT) uses a collection of identical blackboard-based systems to solve problems of monitoring and interpreting data from a set of sensors at spatially distributed locations, covering a region (Lesser et al. 1983). In the MINDS system (Huhns et al. 1987) each user works with a blackboard-based agent to retrieve documents from its local database or to get them from other agents.

In a MAS, the complexity of communications depends upon the agents' characteristics. Reactive agents use sets of communication rules (or activities) that are triggered when specific states hold or predetermined events occur (these events may be messages received from other agents or changes perceived in the environment). The individual message types combined with their expected answers yield a protocol for interaction that spans more than one interaction. Such protocols have been developed for various systems such as the Contract-Net Protocol (Smith Davis 1981) or cooperation protocols (Camarata et al. 1983).

Communications and actions are usually interleaved. Hence, intentional agents must be able to create plans in which both linguistic and non-linguistic actions are combined. Kreifelts and Von Martial (1991) present a negotiation protocol that is used to coordinate the activities of planning agents.

Following Winograd and Flores (1986), we will adopt here a "language/action perspective" where agents are seen as acting through the use of language. Hence, it is relevant to consider speech-act theory (Searle 1969) (Searle et Vanderveken 1985) which provides a theoretical framework for studying the language used by persons when performing various kinds of *speech acts* such as requests, orders, promises. Social agents are able to reason about the mental states (beliefs, intentions, expectations) of other agents. For instance, to satisfy their own goals, they plan their speech acts to affect other agents' beliefs and intentions. Cohen and Perrault (1979) proposed a plan-based theory of speech acts which has been further extended in a theory of relationships among belief, intention and rational communication (Cohen Levesque 1990). Although speech-act theory has been widely applied to natural language processing such as dialogue systems or cooperative response systems (Kaplan 1983), it has not been extensively used to implement communication interactions in MASs: several authors define the messages exchanged by MAS agents as speech acts, but still few systems provide their agents with capabilities to reason on mental states and speech acts.

From a larger perspective, interactions between social agents may be considered as taking place within *conversations*, and conversations may be viewed as coordinated cooperative interactions among agents (Gibbs et al. 1990). Lewis (1969) emphasized the importance of *conventions* (regularities in behaviour evolved by two or more people) to solve recurrent coordination problems. Grice (1975) proposed a theory of cooperation in conversation, considering that each participant in a conversation obeyed a *cooperative principle*: "make your conversational contribution such as required, at the stage it occurs, by the accepted purpose or the direction of the talk exchange in which you are engaged." These theories can provide important foundations to implement sophisticated communication capabilities in MASs.

12.2.5 Multiagent Planning

Bond and Gasser (1988 p. 21) indicate that "we can achieve greater coordination by aligning behaviour of agents toward common goals, with explicit divisions of labour. Techniques such as centralized planning for multiple agents, plan reconciliation, distributed planning, organizational analysis, are ways of helping to align the activities of agents by assigning tasks after reasoning through the consequences of doing those tasks in particular orders." Plan synchronization may be performed at several points: during the plan decomposition, during the plan construction or after the plan construction. Agents' plans may be conflicting because of incompatible states, incompatible orders of activities or incompatible use of resources. Such conflicts can be solved by a particular agent (coordinator or mediator) or a solution may be reached through negotiation (Von Martial 1992).

In *centralized multiagent planning* one agent is responsible for building a multiagent plan that specifies all the agents' planned actions. This approach has been chosen by Cammarata et al. (1983) for the air-traffic control problem: Agents (aircrafts) first negotiate to choose a coordinator among them. The coordinator then creates a multiagent plan to be performed by the aircrafts involved in the conflict. Another way of implementing centralized multiagent planning was proposed by Georgeff (1988): the plans of individual agents are formed first, then a central planning agent collects the local plans and analyzes them to identify potential conflicts. This agent tries to solve the conflicts by modifying the local plans and introducing communication commands into them so that agents synchronize appropriately.

In *distributed multiagent planning*, the planning activities are divided amongst the group of agents. This approach is used when there is not a single agent with a global view of the group activities. Conry et al. (1988) developed a negotiation protocol (called multistage negotiation) for cooperatively resolving resource allocation conflicts. Durfee and Lesser (1987) proposed an approach called *partial global planning* in which agents build local plans and share these plans to identify potential improvements in coordination. Unlike multiagent planning which assumes that a plan is formed before agents begin to act, partial global planning allows agents to interleave planning and actions.

Because multiagent planning requires agents to share and process substantial amounts of information (goals, plans), it also requires more processing resources (computation and communication) from agents than other approaches.

12.3 AN APPROACH FOR DESIGNING MULTIAGENT SYSTEMS FOR HCCW

12.3.1 Methodological Issues

As we have seen in the preceding sections, several interfering issues must be considered when designing a multiagent system. For instance, communication, control, planning and cooperation are activities which are combined together to provide the overall behaviour of an agent in a MAS. The complexity of the design task stems from the multiplicity of issues that must be considered at once by the designer. Significantly enough, the specialized litterature contains very few papers proposing methodological guidelines for developping MASs. Authors usually present new algorithms, knowledge representation schemes and/or agent architectures that are adapted to typical multiagent cases. They do not however describe the methods that should be applied to practical cases in order to develop a MAS using the techniques they propose. Although Decker et al. (1989) do not propose a method for designing MASs, the questions they include in an evaluation questionnaire for multiagent research may help designers considering relevant directions to investigate when starting a MAS project. We think that more efforts should be devoted to methodological issues if multiagent techniques are to be largely applied in organizations. Recent approaches suggest to adapt knowledge acquisition methods such as KADS (Hickman 1989, Heyward et al. 1988) for the development of expert systems using a multiagent framework (Ferraris 1992).

We are not pretending to propose in this chapter a general method for designing MASs. Instead, we would like to illustrate how a design method can be useful to develop multiagent systems for supporting cooperative work (CSCW). Steiner and Mahling (1990) observe that CSCW systems need to support the interaction of humans (and machines) where one of the major hurdles is that of spatial and temporal distribution. They suggest that "viewing human and machine agents as partners in communication, cooperation, problem solving and task execution extends the conventional CSCW paradigm to *"Human-Computer Cooperative Work (HCCW)."* Adopting this view of HCCW, we add that these systems can also be considered as natural extensions of information systems using multiagent techniques. Information system design methodologies have proven to be useful for guiding designers during system development (Olle et al. 1988, Van Assche et al. 1991). The method we propose here can be considered as an extension of such methodologies applied to MASs.

Before presenting the method, let us identify the main characteristics of typical MASs that we are considering for HCCW. In order to simplify the following presentation, we will only consider MASs composed of reactive agents. Although this seems to limit the scope of the approach, reactive agents already provide designers with sophisticated tools to develop HCCW systems. We are interested in group interactions involving persons and machines (Figure 12.1). A group participant may be either an artificial agent or a team grouping a human and an artificial agent: human agents' interactions are mediated by their artificial partners. Artificial agents are linked together by a telecommunication network. Human and artificial agents interact together through communication interfaces supporting

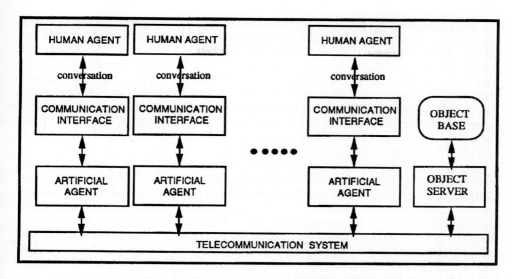

Figure 12.1. A multiagent system for supporting group interactions.

"conversations." The MASs we are considering are heterogeneous by nature: human agents may have different responsibilities and skills, while artificial agents may have varying functionalities and capabilities. Each artificial agent is autonomous: they are able to accomplish various activities by themself and to manage their own local data base. In some situations an agent may require the help of other agents and consequently initiates group activities.

Agents not only interact with one another, but they also have access to common data representing the world in which they evolve. This world contains the artefacts (documents, objects, machines, etc.) that are created or manipulated by agents. We consider a special kind of artificial agent, called object servers. An *object server* manages a data repository, called an *object base*, composed of objects (groups of attributes) that represent parts of the world or artefacts. Depending on the sophistication of the object server, the objects may be structured as a relational database, or as a frame base (with inheritance links, procedural attachments, etc.). An object server is a specialized agent that behaves as a deductive database or a "knowledge service" (Pan et al. 1991).

12.3.2 Group Interactions and Scripts

In human organizations, most group activities are done through conversations which usually conform to pre-established "scripts." From a "language/action perspective" (Winograd and Flores 1986), we consider an extended definition of a "conversation" that corresponds to a script of interactions which take place among a group of agents performing speech acts as well as non-linguistic acts. This perspective leads us to propose a unified model in which communication (speech acts) and action (ordinary acts) are integrated in what we call "communication/action scripts" (Moulin et al. 1991). This approach takes advantage of what has been separately studied up to now in logics of action (Cohen and Levesque 1990) and logics of communication (Searle and Vanderveken 1985).

Individuals play roles in group interactions (Chang 1987). An individual can even play several roles during the same group session. Hence, an agent participating in a group activity should be able to evaluate her role and to select the appropriate plans to play this role. In human organizations the competence issue is most important when group activities have to be organized. When a human agent interacts with a group of agents, she evaluates (often unconciously) other agents' decisions, speech acts and/or activities on the basis of the competence she associates (often subjectively) with each agent within the group. This competence evaluation becomes an important issue when agents have to distribute tasks among group participants. In our approach agents are characterized by their competence to play a given role in a communication/action script.

As an example, let us consider the script consisting of organizing a meeting with the craftsmen of our fictitious workshop. Suppose that each craftsman is helped by an assistant. Craftsman C1 wishes to arrange a meeting with the other craftsmen C2, C3 and C4 to discuss a subject S. He says to his assistant A1: "Dear A1, can you arrange a meeting with C2, C3 and C4? The purpose of the meeting is to discuss subject S. It should take place between dates D1 and D2. We will need about 2 hours for this meeting." Assistant A1 says: "No problem! I will contact their assistants and we will try to arrange the meeting as you request." Assistant A1 phones successively the assistants of craftsmen C2, C3 and C4, in order to find a time period for the meeting which will fit within the schedules of their bosses. The interaction between the assistants implies several exchanges, and negotiations in order to solve the meeting problem. When a time period is found, assistant A1 informs her boss. If no time period is available given the current contraints on the craftsmen's agendas, A1 will inform her boss and ask him if some constraint may be relaxed: changes in dates, in the meeting duration, or in the number of persons involved in the meeting.

If a MAS were implemented in our fictitious workshop, each craftsman would use an artificial agent on a microcomputer as an assistant. In order to solve the meeting problem, artificial agents would interact together, playing roles in a script similar to the scenario involving human assistants that has been described above. Each artificial agent manages the agenda of her craftsman.*

Now, let us examine the negotiation protocol that will be used by artificial agents to arrange the suggested meeting. This protocol, inspired by (Cammarata et al. 1983), aims at determining the most-constrained craftsman, whose agent will be responsible for managing the negotiation. As a matter of fact, the most-constrained agent is best suited for making proposals with a good probability of being accepted, since meeting time slots not fitting in her agenda are not acceptable to the group as a whole. Agents negotiate together, two by two, in the order of their constraint factors. Here is the script description corresponding to this protocol.

One craftsman asks his artificial assistant (AA) to organize the meeting. This AA plays the role of "initiator" of the meeting interactions. In this case the meeting scenario is composed of five phases in which the AAs play different roles: initiation, role determination, negotiation start, negotiation rounds, negotiation termination.

Initiation phase. The initiator-AA contacts the potential-participant-AAs associated with the craftsmen on the meeting list and proposes them that they achieve a "group goal"

---

*Notice that an agenda is divided into days composed of twelve one-hour slots, each slot can take the values "free", "negotiable" (slot reserved for an activity that can be reallocated to another time slot) or "fixed" (slot that cannot be reallocated). All agendas are managed in the same way by artificial agents.

of organizing a meeting S characterized by the relevant parameters. In order to respond to initiator-AA's proposal, each potential-participant-AA contacts her craftsman to ask him if he desires to participate in such a meeting. If her boss accepts, the agent informs the initiator that she commits herself to achieve the "group goal". If her boss does not accept, she informs the initiator that this meeting does not agree with her motivations. Initiator-AA receives the answers sent by the potential-participant-AAs, creates the participant list and sends it to the participant-AAs: this new role is assigned to the potential-participant-AAs who accepted to participate in the meeting organization. For the rest of the negotiation, the initiator-AA is considered as a participant-AA.

Role determination. When she receives the participant list, each participant-AA establishes for her agenda a "constraint factor" which measures her availability during the days (date D1 to date D2) corresponding to the period proposed for the meeting. The constraint factor is computed using the time slots' values for that period.* Each participant-AA sends her constraint factor to every other participant-AA. After the exchange of constraint factors between agents, each participant-AA can establish an ordering of the participant-AAs on an "availability scale" and knows the new roles of participant-AAs: the most-constrained participant (MCP), the second-most-constrained participant (2MCP), the third-most-constrained participant (3MCP), etc.

Negotiation start. The negotiation is led by MCP in order to minimize the number of exchanges between the negotiating AAs. MCP announces the negotiation start to each AA on the participant list and waits for their acknowledgements. Participant-AAs send their acknowledgements to MCP and the negotiation starts.

Negotiation rounds. MCP uses the following strategy: she will first propose to hold the meeting during time periods which correspond to time slots with "free" values in her agenda. The first "free time slot" which is suitable to every participant-AA is chosen. In order to limit communication exchanges, agents negotiate two by two. MCP proposes a meeting slot to 2MCP. If 2MCP cannot accept the proposed slot, she immediately notifies MCP who will have to make another proposal. If 2MCP agrees with the proposal, she transfers it to 3MCP. If 3MCP cannot accept the proposed slot, she immediately notifies 2MCP who transmits the refusal to MCP who will have to make another proposal. This negotiation chain is applied from the most constrained participant to the least constrained participant. If an agreement on the "free time slots" is not reached with other participants, MCP will consider the time slots with the value "negotiable" in her agenda and negotiate with the participant-AAs in order to get an agreement. Proposals are also transmitted along the negotiation chain. When transmitting her answer, each agent also indicates her constraint factor for each proposed period.* MCP memorizes all the constraint factors in order to determine the best compromise for the group of agents.

Negotiation termination. If the negotiation is successful, MCP fixes the meeting date and time and sends the information to each participant-AA who updates her agenda, sends an acknowledgement to MCP and points out the new meeting to her craftsman. If the

---

*Time slots are associated with weights : "free", "negotiable" and "fixed" values correspond respectively to weights 0, 0.5 and 1. The availability factor is calculated as the sum of the weights of the time slots included in the interval [D1, D2], divided by the number of slots included in [D1, D2].

**For an agent, the slots of a proposed period may already be assigned to certain activities. If these activities can be reallocated however, the agent can compute for this period a constraint factor which measures its reallocation cost.

negotiation is unsuccessful, MCP will notify to initiator-AA that no solution is available given the current constraints. Initiator-AA indicates to her craftsman that the negotiation has been unsuccessful and asks if some constraints may be relaxed. If some constraints are relaxed by the craftsman (potential dates for the meeting, potential participants or some time-slot values in his agenda), the negotiation is reactivated at the role-determination phase, given the new meeting parameters.

12.3.3 A Scenario-based Design Approach

There is a strong analogy between a multiagent system composed of agents playing roles and achieving group activities, and a theatrical world in which human actors play their parts according to a given scenario. It is inspiring to examine the way animated movies are created (Culhane 1988). Starting from the initial idea of a scenario (a written script for example), each actor (or character) is quickly sketched according to its physical and psychological characteristics. The story is described through the creation of a "storyboard" which contains series of rough sketches of the salient scenes composing the script. The storyboard is revised until the animation team is satisfied. Then the character designer, the layout artist and the animator cooperate together to draw the series of pictures which will fill in the gaps between the storyboard sketches. Various specialized activities are then undertaken to transform the series of pictures into an animated movie.

We propose a MAS design approach based on the analysis and design of scenarios, called the Multi-Agent Scenario-Based method (MASB method for short).

When managing a MAS project, designers face two possibilities, depending on whether or not a "human MAS" already exists, composed of a group of persons interacting together to achieve specific tasks. When there is no pre-existent "human MAS," designers describe scenarios involving human and artificial agents from scratch.

When analysing an existing MAS, designers describe it in terms of scenarios in which agents play different roles. Usually, an existing MAS is composed of human agents who work together in a human organization (such as an office, a workshop, a team). As we will see in the following sections, different design techniques are used to describe these scenarios and agents' behaviour: behaviour diagrams, data models and transition diagrams.*

Given the models describing an existing MAS composed of human agents, designers can identify existing or new functions that could be assigned to artificial agents. As a result, they can design new scenarios involving human and artificial agents, hence modelling the new MAS. On this basis designers can specify the behaviour of the artificial agents according to the role they play in the scenarios.

The main notions with which we propose to describe scenarios are symbolically presented in Figure 12.2. Human agents (or users) and artificial agents are respectively represented as shaded and non-shaded rectangles. The communication between human and artificial agents takes place within conversational frames.

Agent i is symbolized as a transparent box in order to detail her main components. A role is symbolized by a rectangle with a header indicating the role name. Within the role box, agent activities are also represented as rectangles which can be linked to temporary data contained in the agent's short-term memory (ovals in the role box). An agent can play

---

* Early versions of this method have been described in (Moulin 1983 and 1985). This method has also been successfully applied to the design of information systems and expert systems.

Collaborative Work Based on Multiagent Architectures

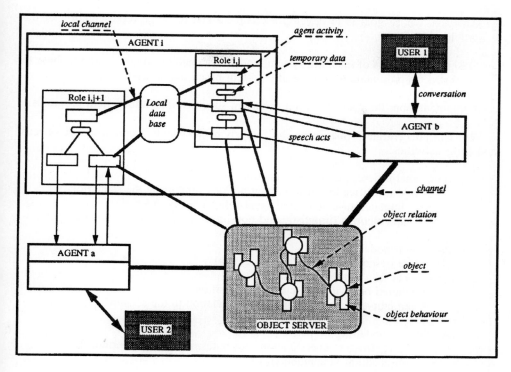

Figure 12.2. Main concepts used to describe scenarios.

several roles. Role activities may use data stored in the local data base (round-corner rectangles within the agent box) through local channels (bold segments linking role activity and the local data base). Channels symbolize the internal agent mechanisms relating her long-term memory (local data base) with the processes supporting her activities. Artificial agents can communicate together by performing speech acts and exchanging messages (represented by arrows).

The world in which agents evolve and act is symbolized by the object server (shaded round-corner rectangle). The object server contains the world knowledge shared by the agents. In Figure 12.2 the object server is represented as a transparent box in order to show its content: objects (circles) associated together by relations (segments linking the circles). Objects are described as sets of attributes and can be associated with specific behaviours (little rectangles overlapped by circles). Hence, objects may be active in the agent world. Note that a MAS may contain several object servers, whenever it is necessary to represent several worlds in which the agents interact. Agents are related to the object server by means of channels which symbolize agents' abilities to perceive the objects of the world as well as their capabilities to act on these objects (by changing their attribute values).

These simple and quite natural notions will be used throughout the method to model scenarios and transform them into system specifications implementing the various components of a MAS.

Like most system development methods, and specifically object-oriented methods (Coad/ Yourdon 1991, Rumbaugh et al. 1991), the MASB method is composed of three phases:

- Analysis, whose goal is to develop a complete and accurate representation of the problem domain;

- Design: the process of mapping system models obtained during the analysis phase to an abstract representation of a specific system-based implementation;

- Implementation: the transformation of design specifications into system programs.

We will concentrate primarily on the analysis phase which is composed of various modelling activities:

    A1 Scenario description (using natural language)
- first identification of important notions supporting the scenario: human/artificial agents, agent roles, objects, agents' interactions (linguistic and non-linguistic acts), object changes, etc.

    A2 Agent modelling
- role description (behaviour diagrams)
- local data modelling
- detailed behaviour description (transition diagrams)
- validation of agent interactions in relation to the scenario

    A3 Object modelling
- object structure specification (attributes, relations, inheritance, procedural attachments)
- object life cycle
- object behaviour
- validation of object/agent interactions in relation to the scenario

    A4 Conversation modelling
- user/agent interactions (connected speech acts, interface specification, etc.)
- validation of conversations in relation to the scenario.

12.3.4 Scenario Modelling

Designers, helped by users, give an initial natural language description of the scenario as in our example in Section 12.3.2. This description emphasizes the roles played by human and artificial agents, the typical information exchanges (or speech acts) and events that occur in the course of the scenario and the actions performed by agents. It is also relevant to describe within the scenario the modifications affecting the world in which agents evolve: object changes and the reactions of objects to changes.

On the basis of this natural language description, designers can use several techniques to model the relevant roles and plans of agents, as well as the characteristics and behaviour of objects of interest in the MAS under analysis. These techniques are presented in the following sections.

In a complex MAS, designers may identify several scenarios. Each scenario is described as mentioned above. If the scenarios are inter-related in some way, the description must emphasize these interelations.

12.4 AGENT MODELLING

12.4.1 Role Description

An agent can play several roles in one or more scenarios. Playing a given role, an agent can perform one or more activities. These activities may either be initiated by the agent or triggered by messages received from other agents (or even by object state changes in the object server). Some situations may induce a role change for a given agent.

Let us come back to the meeting scenario (Section 12.3.2). The craftsmen's assistants can play several roles. During the meeting initiation phase, one agent is the "initiator" and the other agents are "potential-participants." The potential-participants who accept to participate in the meeting become "participants," while those who don't accept are left out. Then the role-determination phase starts. Each participant (including the initiator) plays the role of "role evaluator," computes her constraint factor over the proposed meeting period and sends it to other participants. When all the participants have received all the constraint factors, they are able to order them and then new roles are determined: the "most-constrained participant" (MCP), the "second most-constrained participant" (2MCP), down to the "least constrained participant" (LCP). The negotiation rounds are managed by MCP until the negotiation terminates. In the meeting MAS all the agents have similar behavioural characteristics, but they can play different roles from one meeting scenario to another.

For a given role, a designer can create one or more behaviour diagrams. These diagrams enable a designer to precisely describe the scenario in terms of agent roles and activities (processes), local information or knowledge stores (accumulations) and their contributions (channels) to activities, and interactions (flows) with other agents (environments) playing specific roles. Behaviour diagrams are validated with respect to the natural language scenario description (Section 12.3.4), which may be refined and updated as a result of this validation.

A *behaviour diagram* is a tripartite graph whose nodes are called *processes*, *accumulations* and *environments*. Two types of edges may relate these nodes: processes may be linked to environments by *flows* and processes may be related to accumulations by *channels*. A behaviour diagram is associated to an agent and a role. Processes may eventually be refined in more detailed behaviour diagrams.

Graphically speaking, processes, accumulations and environments are represented respectively by rectangles, round-corner rectangles and double squares. Channels and flows are represented by bold segments and arrows. Processes related to a specific behaviour are embedded in a rectangle which represents the corresponding behavioural plan. The agent's name and role are indicated in the header of this rectangle.

Figure 12.3 presents four behaviour diagrams corresponding to four roles that an agent (called AMSYS for Agenda Management SYStem) can play in the agenda scenario. Let us briefly comment upon these behaviour diagrams (elements of the diagram are indicated in italics in the following presentation.

Diagram 1 describes the behaviour plan *Initiate-meeting-scenario* associated with the *Initiator* role. A *user* (E0) sends the agent a *meeting-request* that is received by process P1.1, which registers the *Meeting-information* (A1): subject, duration, earliest and latest dates, and persons to be contacted. Process P1.2 sends a *meeting-proposal* to the *Potential-participants* (E1) according to the list of invited persons. *Potential participants* (E1) send back their *answers* that are received by process P1.3 which creates the *Participant-list* (A3). When all the agents' answers have been received, the *Initiator* agent sends the *participant-*

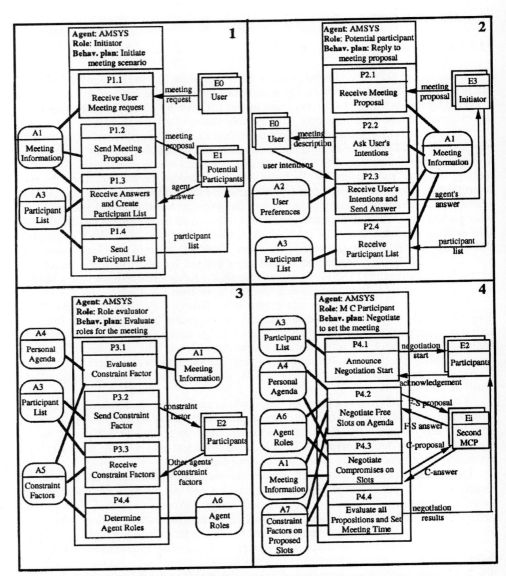

Figure 12.3. Behaviour diagrams for four roles in the meeting scenario.

list to the selected *Potential-participant* agents.

Diagram 2 describes the behaviour plan *Reply-to-meeting-proposal* associated with the P*otential-participant* role. The *meeting-proposal* is received from an agent playing the *initiator* role (E3). Process P2.1 receives the *meeting-proposal* and records it in the *Meeting-information* (A1). Process P2.2 sends the *meeting-description* to the *user* (E0) served by the agent, and asks for her intentions. *User's intentions* are received by process

P2.3, which records them in the *User-preferences* (A2) and sends the *agent's answer* to the *Initiator* agent (E3). If the agent has notified her user acceptance, a *participant-list* will be received by process P2.4, which then records it in accumulation A3.

Diagram 3 describes the behaviour plan *Evaluate-roles-for-the-meeting* associated with the *Role-evaluator* role. This role is played by all agents who accepted to participate in the meeting. The first process (P3.1) evaluates the agent's own *Constraint-factor* (A5) using information stored in the *Meeting-information* (A1) and the agent's *Personal-agenda* (A4). Then, process P3.2 sends the agent's *Constraint-factor* to all the *Participants* (E2) indicated in the *Participant-list* (A3). Process P3.3 receives the *constraint-factors* from the *Participants* (E2) and stores it in accumulation A5. When all *Constraint-factors* have been received, process P4.4 determines the *Agent-roles* (A6) for each agent participating in the meeting.

Diagram 4 describes the behaviour plan *Negotiate-to-set-meeting* associated with the *Most-constrained-participant* role (MCP role). Process P4.1 announces the *negotiation-start* to all *Participants* (E2) listed on the *Participant-list* (A3), and waits for *acknowledgements* from the contacted agents. Process P4.2 makes free-slot proposals (*F-S proposals*) for the meeting on the basis of the *Meeting-information* (A1) and available time slots contained in the MCP agent's *personal agenda* (A4). These *F-S proposals* are sent to the *Second-MCP (Ei)* as stored in the *Agent-roles* accumulation (A6). *F-S answers* are also received from *Second-MCP*. Process P4.2 generates proposals until one of them is accepted by all *Participants* (this information is transmitted from one agent to another along the negotiation chain as described in Section 3.2). If no F-S proposal is accepted, process P4.3 generates a compromise proposal (*C proposal*) for each slot on the *Personal agenda* (A4) for which a compromise can be made, and sends it to *Second MCP*. *Second MCP* transmits this *C-proposal* to other agents along the negotiation chain and transmits back their answers, which are received by process P4.3 of MCP. Process P4.3 records in accumulation A7 the *Participants' Constraint-factors* for each proposed slot. Then, process P4 evaluates all propositions, computes the best one (which minimizes the sum of participants' constraint factors over a given period) and sends the *results* to all the *Participants* (E2).

Diagram 4 is an example of a high-level behaviour plan: some of its processes (P4.2 and P4.3) could be further refined in new behaviour plans.

Due to lack of space, other roles are not described in this paper. The displayed diagrams permit the reader to understand how behaviour diagrams are used to model a scenario and to check them against the natural language scenario description obtained during the first phase of the method.

Note that behaviour diagrams may be considered as system plans (Moulin 1983) in a similar way as data-flow diagrams are used in information system design methodologies to represent the architecture of an information system (Moulin 1985). In fact, when we model an agent's behaviour in the context of a given scenario, we can consider this agent as a system which interacts with other systems (other agents). Hence, system modelling techniques apply at this level of abstraction. Behaviour diagrams will be used to further refine the agent characteristics in terms of local data structures and transition diagrams which represent in a detailed way the agent behaviour.

12.4.2 Local Data Modelling

In the context of a given scenario, a designer creates a set of behaviour diagrams for each agent. The set of accumulations contained in these behaviour diagrams corresponds to the data and knowledge used by the agent. During the local data modelling phase, a designer analyses the contents of these accumulations in terms of attributes and integrates these attributes into the agent's local data conceptual structure.

This data conceptual structure may be represented respectively by "entities and relationships," by "classes, sub-classes and relationships" or by "frames," depending on whether the agent's local data base is implemented with a relational data base, an object-oriented data base or a frame base. In order to create the data conceptual structure, a designer may choose any data-modelling approach that she prefers (Chen 1976, Moulin 1983, Olle et al. 1988, Coad/Yourdon 1991, Rumbaugh et al. 1991). For our example (Figure 12.4) we use an entity-relationship model with binary relationships (entities are represented by rectangles and relationships by ovals, cardinalities are indicated on edges) to represent the agent's local data conceptual structure.

The entity *MEETING* contains the meeting information: Meeting-id, the Initiator-name, the Subject and Duration, the Earliest and Latest dates proposed by the user, the Accepted-date and Period are instantiated when the metting is set. The Meeting-status attribute will be discussed later. The entity *PERSON* records the information about the other persons known by the agent: Person-id, -name, -address, -phone and -job. The entity *PERIOD* records the time slots assigned to activities in the agent's agenda. Its attributes are Period-id, the Date, the Begin- and End-time (corresponding to time slots covered by the period), the Assigned-activity for the agent and the Activity-status (with values "fixed" or "negotiable"). The relation *Potentially-participates* is used by the initiator-agent to record the list of persons to be contacted, and the Acceptance-status attribute records the fact that a person has accepted (or not) to participate in the meeting. For each agent agreeing to participate in the meeting, the initiator creates an instance of the entity *PARTICIPANT*, with a Participant-id and a Role attribute which is used to record the role of the participant when it is determined. The *Initiator* relates this instance of *PARTICIPANT* to the corresponding *MEETING* and *PERSON* instances, using the relations *Involves* and *Corresponds-to*. The initiator sends all participant-agents a copy of the participant list (the list of participant-ids for the *MEETING* and instances of the *Corresponds-to* relation together with the person-id), and each participant-agent creates the corresponding instances of the *PARTICIPANT* entity in his local data base. The agents then compute their constraint factors and send them to the other participants. Each agent records these constraint factors in the Global-constraint-factor attribute of her *PARTICIPANT* entity. The most-constrained participant (MCP) generates proposals that she records in her local data base using the *Potentially-assigned* relation, whose Assignation-status attribute records the proposal status. When she gets the answers back from other participants, the MCP agent creates the relevant instances of the relation *Occupied* and eventually registers the Period-constraint-factor for the appropriate period and the appropriate participant.

The conceptual data structure specifies the static properties of the agent's local data base. At any given time the state of an agent is characterized by the content of its local data base. The agent's evolution is reflected by the changes that occur in the attribute values. Not all state transitions are allowed, however. For example the "proposed" meeting must be "negotiated" before eventually being "abandoned;" thus, the state change between "proposed" meeting and "abandoned" meeting is not allowed. In the MASB method, we can specify the allowed state transitions for entities (or objects) of the agent's local data conceptual structure by means of life-cycles for those entities (Moulin 1985, Creasy Moulin 1991).

An *entity life-cycle* (resp. object life-cycle) diagram is a connected graph in which the

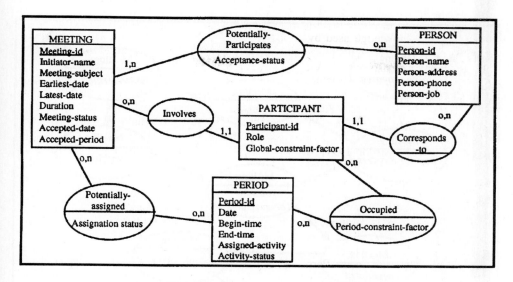

Figure 12.4. Agent local data base schema.

nodes represent the entity (resp. object) states, and the edges represent the transitions allowed between these states. The entity (resp. object) life-cycle indicates which changes are allowed in the instances of a given entity (resp. object). Figure 12.5 shows the life-cycle associated with the entity *MEETING*. This life-cycle diagram specifies all the states through which a *MEETING* instance can evolve, whatever role the agent plays. The names of the states are indicated in italic in the ovals. This life-cycle indicates the various states through which a meeting entity evolves in an agent's local data base. In the initiator-agent data base the meeting is created in the "to-be-proposed" state and becomes "wait-for-answers" until all the potential participants' answers have been received. The symbol |-> indicates the creation of an instance in the corresponding state. If all the potential participants refuse to participate, the meeting instance becomes "refused." In a potential-participant local base the meeting instance is created in the state "to-be-verified," and becomes "verified" (resp. "refused") if the user accepts (resp. refuses) to participate in the meeting. Then the meeting instance takes the states "roles-to-be-evaluated," "negotiated," "compromised," "set," or "abandonned" depending on the negotiation stage. The symbol ->| indicates that the instance can be deleted from the corresponding state. In the agent's local data base the instance state of an entity (or object) is recorded using a special attribute called entity (or object) status. In Figure 12.4, we included the Meeting-status attribute in the *MEETING* entity.

Using entity (or object) life-cycles, a designer conceptually specifies the allowed state transition for the entity (or object) on the basis of semantic properties of the entity (or object). They will be used in the next design step to further refine the specification of the agent's behaviour, using transition diagrams.

12.4.3 Detailed Behaviour Description

Behaviour diagrams provide global models of an agent's roles and activities in the

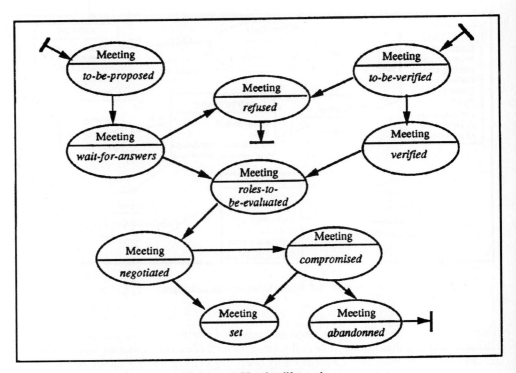

Figure 12.5. Meeting life-cycle.

context of a given scenario. The local data base structure describes the data manipulated by the agent and entity life-cycles indicate relevant state changes in this local data base. The last step of the analysis phase combines and further refines these specifications in order to provide a detailed specification of the agent's behaviour in the form of transition diagrams.

A *transition diagram* is a detailed specification corresponding to the last refinement level of a process included in a behaviour diagram. Transition diagrams correspond to finite-state machines that completely specify in detail teh behaviour of an agent at a conceptual level, and represent the state changes related to the agents' behaviour, each state change being specified by a transitory group.

A *transitory group* is composed of a set of triggering elements, a transition and a set of triggered effects. A transition is an agent function (procedure or activity) that changes the agent's state (by modifying her local data base), sends messages to other agents or accesses data in an object-server.

A transition can be triggered by a combination (disjunction and/or conjunction) of elements and can generate a combination (disjunction and/or conjunction) of triggered effects.

In Figure 12.6 the transition is symbolically represented by a rectangle named "transition." The triggering elements are represented at the origin of the arrows pointing towards the transition rectangle, and the triggered effects are represented at the destination of the arrows originating from the transition rectangle.

Let us consider the triggering elements, starting from the left (upper part of

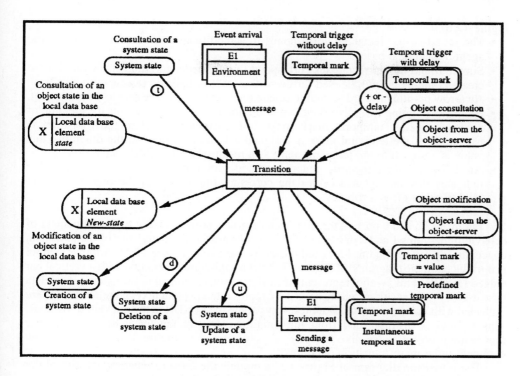

Figure 12.6. Triggering and triggered elements associated with a transition.

Figure 12.6).
- A round-cornered rectangle with a vertical bar represents a local data base entity (or object) state, the state name is specified in italics; this element corresponds to a consultation of that state.
- Intermediate system states are represented by simple round-corner rectangles. If a "t" is specified on the arrow, the state is considered as transitory and deleted after the triggering of the transition. If there is no "t" on the arrow, the system state is considered as persistent and is not deleted after the triggering of the transition.
- The double rectangle represents another agent in the agent's environment. The transition receives a message from the environment, which can also be considered as an external event from the agent's point of view.
- A double round-corner rectangle represents a temporal mark. A temporal mark activates the transition when the current time equals the time specified on the temporal mark. In addition, we can indicate in an associated circle the addition (symbol +) or subtraction (symbol -) of a delay to the temporal mark. Temporal marks enable the designer to relate the agent's behaviour to specific dates according to the time-coordinate system of the multiagent system.
- A double round-corner rectangle with a vertical bar specifies a consultation of an object from the object server, hence modelling the agent's perception of her environment. Next we will comment upon the triggered effects, starting from the left (lower part of

Figure 12.6).
- The round-corner rectangle corresponds to a modification of an object state in the agent's local data base.
- Simple round-corner rectangles correspond to intermediate system states that can be created (simple arrow), deleted (arrow with a "d") or updated (arrow with a "u").
- An arrow pointing towards a double square represents a message sent to another agent.
- An arrow pointing towards a double round-corner rectangle represents the creation of an instantaneous temporal mark (using the current time when the transition is triggered). If the temporal mark is assigned a value ("=value" in the diagram), this value is assigned to the temporal mark as a result of the transition activation.
- An arrow pointing towards a double round-corner rectangle with a vertical bar indicates that the agent accesses the corresponding object in the object-server to modify some of its attributes.

Figure 12.7a and b present the transition diagrams corresponding to the refinement of behaviour diagrams 1 and 2 of Figure 12.3, considering the characteristics of the local data conceptual structure (Figure 12.4) and the entity life-cycle (Figure 12.5). Another symbol is introduced in the diagrams of Figure 12.7: the round-corner rectangle with a vertical bar tagged Loc DB. This symbol represents the information contained in the agent's local data base used by the transitions: entity names are represented in plain text and relationships are indicated with italic characters. Double arrows relate the Loc DB to the relevant transitions. This symbol is distinguished from the round-cornered rectangles with a vertical bar representing object states, like Meeting *to-be-proposed* in Figure 12.7a. We will now comment upon these transition diagrams.

The transition diagram of Figure 12.7a models the *Initiate-meeting* activity of the *Initiator* role and provides the detailed specification of the corresponding behaviour plan (Figure 12.3.1).

The first transition, *Receive-recommendation*, is triggered by a *meeting-request* message received from the *user* (E0). The corresponding procedure, *Receive-user-meeting-request* creates a *Meeting* instance in the local data base as well as instances of the relation *Potentially-participates* which links the *Meeting* instance to the relevant *Person* instances. The effect of this transition is to set the *Meeting status* to the value "to-be-proposed."

The second transition, *Transmit-request*, is triggered by the "to-be-proposed" state of the *Meeting* entity. The corresponding procedure, *Transmit-request*, is repetitively executed for all potential participants: repetition is indicated as "*f.a. potential participant*" at the bottom part of the transition rectangle. For each potential participant found in the local data base (according to the *potentially-participates* relation), the procedure sends a *meeting-proposal* message to the corresponding agent (E1). When all potential participants have been considered (i.e. the end-condition of the repetition is reached), the intermediate state *Receive-answers* is created (indicated on the arrow by a symbol "e," meaning "end") and the *Meeting-status* of the *Meeting* instance becomes "wait-for-answers."

The third transition, *Receive-answers*, is triggered by the conjunction (represented by the arrows linking segments) of the *Receive-answers* intermediate state and messages received from *Potential-participants* (E2) which indicate either a *Refusal* or an *Acceptance*. The corresponding procedure, *Receive-anwsers-and-check-if-all-have-been-received*, updates the *Acceptance-status* of the *Potentially-participates* relation in the local data base, and eventually creates an instance of the *Participant* entity and instances of the associated relations: *involves* and *corresponds-to*. This procedure checks if all the answers have been

Collaborative Work Based on Multiagent Architectures 285

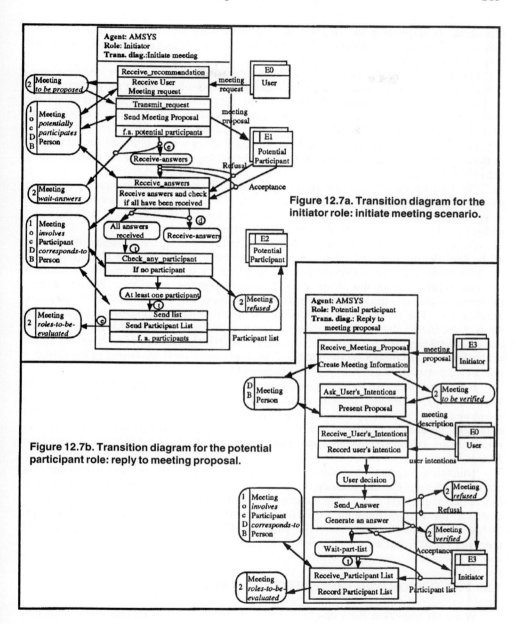

Figure 12.7a. Transition diagram for the initiator role: initiate meeting scenario.

Figure 12.7b. Transition diagram for the potential participant role: reply to meeting proposal.

received: when the last answer is received it deletes the *Receive-answers* intermediate state (arrow with a symbol "d") and creates an intermediate state *All-answers-received*.

The fourth transition, *Check-any-participant*, is triggered by the transitory state *All-answers-received* (which is deleted after the execution of the transition). The corresponding procedure, *If-no-participant*, checks the local data base and either assigns the value "refused" to the *Meeting-status* of the *Meeting* instance, or creates an intermediate state *At-least-one-participant*.

The fifth transition, *Send-list*, is triggered by the transitory state *At-least-one-participant*. The corresponding procedure, *Send-participant-list*, creates the participant list on the basis of the information contained in the local data base and sends it to *all the participants* (E2). At the end of this repetitive procedure, the *Meeting* instance state becomes "roles-to-be-evaluated."

The transition diagram of Figure 12.7b models the *Reply-to-meeting-proposal* activity of the *Potential-participant* role and provides a detailed specification of the corresponding behaviour plan (Figure 12.3.2).

The first transition, *Receive-meeting-proposal*, is triggered by a *meeting-proposal* message received from the *Initiator* agent (E3). The corresponding procedure, *Create-meeting-information*, creates an instance of the *Meeting* entity in the local data base and assigns the value "to-be-verified" to its *Meeting-status*.

The second transition, *Ask-user's-intentions*, is triggered by the "to-be-verified" state of the *Meeting* instance and sends a *meeting-description* message to the *User* (E0).

The third transition, *Receive-user's-intentions*, is triggered by a *user-intentions* message received from the *User* (E0). The corresponding procedure, *Record-user's-intention*, creates an intermediate state *User's-decision* which triggers the fourth transition, *Send-answer*. The corresponding procedure, *Generate-an-answer*, can be executed in two ways: if the *User's-decision* is negative, the status of the *Meeting* instance is set to "refused" and a *Refusal* message is sent to the *Initiator*; if the *User's-decision* is positive, the status of the *Meeting* instance is set to "verified," an *Acceptance* message is sent to the *Initiator* and an intermediate state *Wait-part-list* is created.

The fifth transition, *Receive-participant-list*, is triggered by the conjunction of the *Wait-part-list* intermediate state and a *Participant-list* message received from the *Initiator* (E3). The corresponding procedure, *Record-participant-list*, updates the local data base, creating instances of the *Participant* entity and *involves* and *corresponds-to* relationships. It also assigns the value "roles-to-be-evaluated" to the *Meeting-status* of the *Meeting* instance.

Transition diagrams provide a detailed specification of the agents' behaviour at a conceptual level. The consistency of these diagrams can be checked by looking at the behaviour plans, data conceptual structure and entity life-cycles. If any inconsistency is detected, the analyst finds the erroneous elements in these models and corrects them. Hence, the modelling process is an iterative one which aims at obtaining a complete and consistent specification for each role of each agent.

12.5 THE DESIGN AND IMPLEMENTATION PHASES

When all transition diagrams have been obtained, the analysis phase is completed. During the design phase designers transform the agents' transition diagrams and data conceptual structures into specifications adapted to the programming language or development tools used for system implementation. If an object-oriented language is used, the data conceptual structures are specified in terms of classes and specific relations (inheritance, aggregation, etc.),whereas transition diagrams are transformed into methods assigned to the relevant object classes. If a blackboard system is used for implementing agents, the data conceptual structure is used to specify blackboard data components whereas transition diagrams are transformed into knowledge sources.

We developed SMAUL1 (Cloutier Moulin 1993) a tool, which enables a designer to declaratively describe agents in terms of roles, transition diagrams and data conceptual structures. The tool automatically generates the program structures of the whole multiagent system in the Prolog language. The programmer only needs to add the specific code for each procedure corresponding to the transitions of transition diagrams.

12.6 OBJECT MODELLING

In Section 12.3.3 we introduced the notion of object server used to model objects contained in the world in which agents evolve and to simulate their behaviour. Since the meeting scenario mainly deals with agent's interactions and role changes, it does not provide an opportunity to illustrate the use of the object server. In this section we will briefly describe a new scenario which illustrates how the control of a MAS can be shared between objects and agents.

12.6.1 Scenario Description

Let us recall that our four craftsmen create various kinds of guitars (acoustic, electric, etc.). In the present scenario we will simulate the organization of a flexible workshop. Craftsmen work in different rooms linked by a circular conveyor belt on which guitars are conveyed from one transformation post to the other. Each craftsman knows the operations he has to execute on each type of guitar, but he does not know where a given guitar should be sent after he has completed his operations. Depending on their characteristics, the guitars have to visit the craftsmen's rooms in a different order.

For example, Figure 12.8 presents the path that must be followed by electric guitars in the workshop: the internal order is issued by the administrator; then the guitar goes (1) to the assembler who assembles it; then, it is sent (2) to the electronics specialist who installs the electrical system; then it goes (3) to the fine-work specialist who applies the finishing touches to it; then it goes to the electronics specialist who tests it; finally, it goes back (5) to the administrator for delivery. Guitars of another type may have a different path in the workshop. In this flexible workhop we consider that guitars are active objects knowing the transformation process they must go through: guitar instances may take different states, and depending on their current state they must visit a

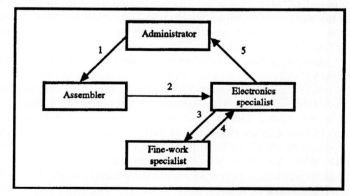

Figure 12.8. Electric guitar transformation in the workshop.

specific room in the workshop to undergo a given transformation. Each guitar type is associated with a behaviour plan, which records the path that should be followed by instances of this type in the workshop.

We suppose that the workshop is managed by a MAS: each craftsman possesses an electronic assistant (agent) who is able to communicate with other agents. The object server is a special agent who records information about objects manipulated in the workshop: she is responsible for managing the data contained in object instances and for activating objects' behaviour plans when appropriate. For each guitar type, the object server contains an object-class describing the guitar's characteristics as well as its behaviour. When an order is received by the administrator, his assistant agent asks the object server to create a new guitar instance in the object base corresponding to the type ordered by the customer. This guitar instance becomes active. On the basis of its behaviour plan and its evolving state, it will direct the pallet containing the physical guitar towards the appropriate craftsmen's posts along the conveyor belt.

In the following sections we present how the MASB method can be used to model object behaviour (corresponding to step A3 of the method presented in Section 12.3.3).

12.6.2 Object Structure and Life-cycles

In Section 12.4.2 we presented an approach permitting a designer to model the structure of an agent's local data base: static properties of an object were modelled with entity relationship diagrams and dynamic properties with entity life-cycles. However, an agent's local base only contains passive objects: there is no behaviour attached to these objects and their dynamic properties only specify which state transitions are allowed for object instances. These transitions are effectively imbedded in the agent's transition diagrams which modify the local data base content and hence completely control its objects.

Objects contained in the object server may be active: we can attach behaviour plans to them, activated by the object server when specific conditions are reached.

The approach we propose for modelling active objects of the object server is an extension of the techniques used for modelling an agent's local data base.

After analysing the natural language description of the scenario, an analyst applies a data-modelling technique (such as an entity-relationship approach or an object-oriented approach) to obtain the static data conceptual structure: object attributes, relationships, etc. In the workshop example, we obtain a class *GUITAR* (with attributes: Guitar-id, Guitar-type, Guitar-components, etc.) as well as sub-classes such as *ELECTRIC-GUITAR*, *ACOUSTIC-GUITARE*.

An analyst then studies the scenario description in order to characterize the behaviours of the various objects. In a first step, the behaviour of an object can be modelled using object life-cycle diagrams. Figure 12.9 presents the life-cycle diagram corresponding to the *Electric-guitar* object. The characteristics of object life-cycles have already been discussed in Section 12.4.2. The only difference we must mention stems from the fact that an object may be active in the object server. Hence, some object state changes are controlled by the object itself (resulting from transitions contained in the object's behaviour plans), while other state changes are not controlled by the object (resulting from agents' modifications in the object base). This difference is shown in the object life-cycle diagram. Object states are represented as round-corner rectangles with black edges. State transitions under the control of an object are represented by rectangles, while uncontrolled transitions are represented by simple arrows linking the relevant states.

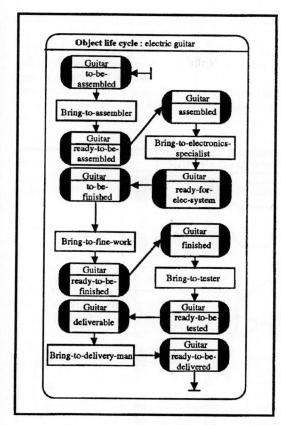

Figure 12.9. Electric guitar life cycle.

For example, an instance of an electric guitar is created in the state "to-be-assembled" which can activate a transition controlled by the *GUITAR* object: *Bring-to-assembler*. This transition changes the state of the instance to "ready-to-be-assembled." From this state there is a transition which transforms the object state to "assembled." This transition is not under the control of the *GUITAR* object and is represented by a simple arrow: it depends on an agent's activity (the assembler in this case).

The object life-cycle of Figure 12.9 presents the complete set of allowed state changes that can affect an instance of an *ELECTRIC GUITAR*. If necessary, an analyst can anotate transitions to give more details about any condition applying to them.

12.6.3 Object Behaviour

Active objects contained in an object server can be modelled as "finite-state machines". Hence, object's behaviour can be specified in a similar way to an agent's behaviour, using a special kind of transition diagram called an object behaviour diagram.

An *object behaviour diagram* is a restricted form of transition diagram (see Section 12.4.3):

- Triggering elements are object states represented as in the object life-cycle diagram;
- Triggered effects are either object states or messages that can be sent to agents;
- Transitions are associated to procedures which are handled by the object server;
- The states appearing in an object behaviour diagram may correspond to any object contained in the object server.

These restrictions follow from the objects' properties. Since objects can interact together, states associated with different objects can appear in the same object behaviour diagram. Objects can send messages to agents in order to inform them about world changes.

Figure 12.10 presents the behaviour diagram of the *ELECTRIC GUITAR* object. It contains the transitions: *Bring-to-assembler*, *Bring-to-electronics-specialist*, *Bring-to-fine-work*, *Bring-to-tester*, *Bring-to-delivery-man*. We notice that there is an *ELECTRIC GUITAR* object state as a triggering element for each transition and that triggered effects consist in an object state and a message being sent to the relevant agent. The characteristics of this particular diagram follow from the fact that the behaviour of this object comprises the knowledge of paths that the object must follow in the workshop.

From a methodological point of view, object behaviour diagrams appear as a further refinement of object life-cycle diagrams: transitions of the object life-cycle that are under the object control (represented as rectangles in Figure 12.9) become the transitions in the object transition diagram, which is completed with the relevant object states and messages as triggering elements and triggered effects.

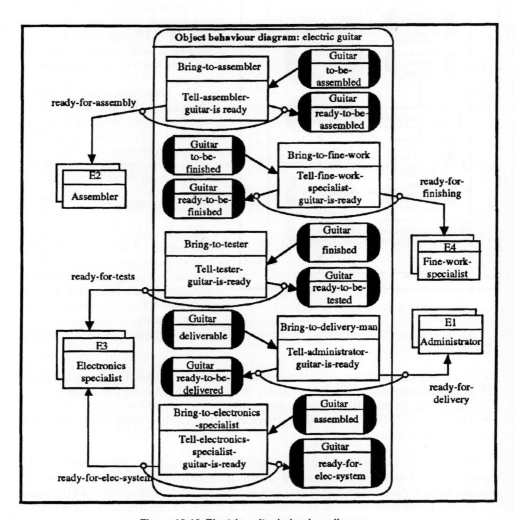

Figure 12.10. Electric guitar behaviour diagram.

12.6.4 Object/Agent Interaction Validation

Since objects and agents may interact, these interactions must be validated on the basis of the scenario description. They must also be mutually checked in order to verify that the behaviour of agents and objects are consistent.

Figure 12.11 presents the transition diagram *Install-electric-system* of the role *Electric-system-manipulator* for the *Electronics-specialist* agent. We note that the transition *Receive-Electric-guitar* is triggered by a message *ready-for-elec-system* sent by the *ELECTRIC-GUITAR* object. This transition creates a system state *Wait-for-electric-system* and sends a message *request-an-electric-system* to the *Administrator* who manages the stock of electrical devices.

The second transition *Work-on-guitar* simulates the work done by the craftsman on the guitar. It is triggered by the message *electric-system-available* sent by the *Administrator*, in conjunction with the transitory state *Wait-for-electric-system* and the consultation of the *Guitar data* in the object server. The effect of this transition is to update the *Guitar-status* (= *"to-be-finished"*) in the object server.

Although this example is quite simple, it illustrates how objects' and agents' behaviour can be mutually checked for consistency.

12.6.5 Object Structure and Behaviour Design and Implementation

The design and implementation of objects' structures and behaviours are done during the design and implementation phases as indicated in Section 12.5. If an object-oriented language is used, the conceptual structures of an object are specified in terms of classes and

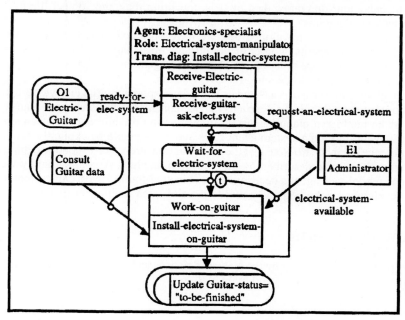

Figure 12.11. Install-electrical-system Behaviour Diagram.

specific relations (inheritance, aggregation, etc.),whereas object behaviour diagrams are transformed into methods assigned to the relevant object classes. If a blackboard system is used, the conceptual structure of an object is used to specify blackboard data components whereas object behaviour diagrams are transformed into knowledge sources. The SMAUL1 tool (Cloutier Moulin 1993) provides utilities to declaratively describe conceptual structures of objects and behaviour diagrams. The tool automatically generates the program structures that are integrated into the multiagent system. The programmer only needs to add the specific code for each procedure corresponding to the transitions of object behaviour diagrams.

12.7 CONCLUSION

In this chapter we described the Multi-Agent Scenario-Based method, a multiagent system design approach based on the analysis and design of scenarios. This method is well suited to model systems supporting human computer cooperation work (HCCW). We are particularly interested in group interactions involving persons and machines where the artificial agents are autonomous: they are able to accomplish various activities by themselves and to manage their local data bases. In certain situations an agent may require the help of other agents and consequently initiating group activities. Agents do not only interact together, but they also have access to common data representing the world in which they evolve. These data are managed by an agent called the object server which contains active objects.

Different design techniques have been proposed to describe these scenarios and agents' behaviour: behaviour diagrams, data models, transition diagrams, object life-cycles and object behaviour diagrams.

In Section 12.3.1 we mentioned that we have only considered multiagent systems composed of reactive agents in order to simplify the presentation of the method. Altough this seems to limit the scope of the approach, reactive agents already provide designers with sophisticated design tools to develop HCCW systems.

This method can be used as a foundation for developping multiagent sytems composed of intentional agents. Behaviour diagrams may be considered as stereotyped plans that can be used by an intelligent planning system implementing the inference mechanisms of an intentional agent. In such a system agent roles and goals would be explicitly manipulated by the inference mechanism. Note that the names of transitions correspond to detailed goals associated with the plan described by the transition diagram. A goal hierarchy would be specified to model each agent's decision space (Lizotte Moulin 1990). Our current research aims at developping a method and a tool to develop such multiagent systems.

ACKNOWLEDGEMENTS

This research is supported by the Natural Sciences and Engineering Research Council of Canada (grant OGP 05518) and by FCAR.

REFERENCES

1. ACM (1991) Communications of the ACM, *Special Issue on Computer Supported Collaborative Work,* vol 34, No.12 (December, 1991).
2. Allen, J. F., H. A. Kautz, R. N. Pelavin, and J.D. Tenenberg. *Reasoning about Plans.* San Mateo, CA., Morgan Kaufmann(1991).
3. Adler, M.N., A. B. Davis, R. Weihmayer, and R. W. Worrest. *Conflict-resolution strategies for non-hierarchical distributed agents.* Gasser, Huhns (1989), pp. 139-161.
4. Bond, A. H. "The cooperation of experts in engineering design." In (Gasser, Huhns 1989), pp 463-484.
5. Bond, A.H., and L. Gasser, Eds. *Readings in Distributed Artificial Intelligence.* Morgan Kaufmann(1988).
6. Cammarata, S., D. McArthur, and R. Steeb. "Strategies of Cooperation in Distributed Problem Solving." *Proc. 8th Joint Conf. on AI.* Karlsuhe, West Germany (1983), pp. 767-770.
7. Chaib-draa, B., B. Moulin, and R. Mandiau. "Trends in Distributed Artificial Intelligence." *Artificial Intelligence Review.* N 6, Millot (1992), pp. 35-66.
8. Chang, E. *Participant systems for cooperative work.* Huhns (1987), pp. 311- 339.
9. Chen P. P. "The entity-relationship model: Toward a unified view of data." *ACM Transactions on Database Systems* 2 (1976), pp. 9-36.
10. Cloutier, L., and B. Moulin. "L'environnement de conception de SMAUL1: un ensemble d'outils pour le développement de systèmes multiagents." *The Proceedings of ICO'93 Conference.* Montréal (May, 1993).
11. Coad P., and E.Yourdon. *Object-Oriented Analysis.* Prentice-Hall (1991).
12. Cohen, P.R., and H. Levesque. "Rational interaction as the basis for communication." In P. R. Cohen, J. Morgan, and M. E. Pollack, Editors, *Intentions in Communication.* MIT Press (1990).
13. Cohen, P.R., and R. Perrault. "Elements of a plan-based theory of speech acts." *Cognitive Science.* N. 3 (1979), pp. 177-212.
14. Conry, S.E., R. Meyer, and V. R. Lesser. "Multistage negotiation in distributed planning." Bond Gasser (1988), pp. 367-364.
15. Creasy P., and B. Moulin. "Approaches to data conceptual modelling using conceptual graphs, in proceedings of the Sixth Annual Workshop on Conceptual Graphs." SUNY, Binghamton, NY (July, 1991).

16. Culhane, S. *Animation, from Script to Screen*. Colombus Book London(1988).

17. Decker K. S., E. H. Durfee, and V. R. Lesser V. R. *Evaluating research in cooperative distributed problem solving.* Gasser Huhns (1989), pp. 485-519.

18. Demazeau, Y., and J-P Müller. *Decentralized Artificial Intelligence.* Elsevier Pub. (1990).

19. Demazeau, Y., and J-P Müller. *Decentralized Artificial Intelligence,* **2**. Elsevier Pub. (1990).

20. Durfee, E. H., V. R. Lesser, and D. D. Corkill. "Trends in Cooperative Distributed Problem Solving." *IEEE Transactions on Knowledge and Data Engineering*, Vol.1 No.1 (March, 1989), pp. 63-83.

21. Durfee, E. H., and V. R. Lesser. "Using partial global plans to coordinate problem solvers." *Proceedings of IJCAI'87* (1987), pp. 875-883.

22. Ferraris, C. "Acquisition des connaissances et raisonnement dans un univers multi-agents: application à la prise de décision en génie civil urbain." Thèse de doctorat, Université de Nancy 1(1992).

23. Gasser, L., and M. N. Huhns, Eds. *Distributed Artificial Intelligence* (Volume 2), Morgan Kaufmann (1989).

24. Georgeff, M. (1988), Communication and interaction in multiagent planning, in (Bond Gasser 1988) pp. 200-204.

25. Greif, I., editor. *Computer-supported cooperative work: a book of readings*. San Mateo, CA, Morgan Kaufmann (1988).

26. Gibbs, R. W., and R.A.G. Mueller. "Conversation as coordinated cooperative interaction." Zachary et al. (1990), pp. 95-114.

27. Grice, H. P. "Logic and conversation." P. Cole and J. Morgan, Eds., *Syntax and Semantics 3: Speech Acts*, New York, Academic Press (1975).

28. Heyward S.A., B.J. Wielinga, and J.A. Breuker. "Structured analysis of knowledge." J. Boose and B. Gaines, Eds., *Knowledge Acquisition Tools for Expert Systems.* Academic Press (1988), pp. 149-160.

29. Hickman, F. *Analysis for Knowledge-Based Systems: a Practical Guide to the KADS Methodology*. Ellis Horwood (1989).

30. Huhns, M. N., Eds. *Distributed Artificial Intelligence.* Morgan Kaufmann (1987).

31. Huhns, M. N., U. Makhopadhyay, L.M. Stephens, and R.D. Bonnell. *DAI for document retrieval: the MINDS Project.* Huhns (1987), pp. 249-283.

32. Kaplan S. J. "Cooperative responses from a portable natural language database query system." M. Bradie and R. C. Berwick, Eds., *Computational Models of Discourse*, MIT Press.

33. Kreifelts, T., and F. Von Martial. "A negotiation framework for autonomous agents." Demazeau Müller (1991), pp. 71-88.

34. Lesser, V. R., and D.D. Corkhill. "The distributed vehicle monitoring testbed: a tool for investigating distributed problem solving networks." *AI Magazine*. (Fall 1983), pp. 15-33.

35. Lewis, D. *Convention*. Cambridge MA, Harvard University Press (1969).

36. Lizotte, M., and B. Moulin. "A temporal planner for modelling autonomous agents." Demazeau, Müller (1990), pp. 121-136.

37. Malone, T.W. "Organizing information processing systems: parallels between human organizations and computer systems." Zachary et al. (1990), pp. 56-83.

38. Mintzberg, H. *The Structuring of Organizations*. Englewoods Cliffs, NJ, Prentice Hall (1979).

39. Moulin, B. "The use of EPAS/IPSO approach for integrating Entity Relationship concepts and Software Engineering techniques." C.G. Davis, S. Jajodia, P.A. Ng, R. Yeh, Editors, *Entity-Relationship Approach to Software Engineering*, Proceedings of the 3rd International Conference on Entity-Relationship Approach, Anaheim CA. (1983), North Holland.

40. Moulin, B. "La Méthode E.P.A.S.: plans de systèmes et modélisation conceptuelle des données, cycles de vie, diagrammes de transition." *Rapport de recherche DIUL-RR 8507 à 8509*, Université Laval (Septembre 1985).

41. Moulin, B., B. Chaib-Draa, and L. Cloutier. "A multi-agent system supporting cooperative work done by persons and machines." *Proceedings of the 1991 IEEE International Conference on Systems, Man and Cybernetics*, Charlottesville, VA, (October, 1991).

42. Nii, H.P. "Blackboard Systems : The blackboard model of problem-solving and the evolution of blackboard architectures." *AI Magazine*, vol. 7(3), (1986), pp. 39-53.

43. Olle, T. W., H. G. Sol, A.A.Verrijn-Stuart, J. Hagelstein, I.G. Macdonald, C. Rolland, and F.J.M. Van Assche. *Information Systems Methodologies: A Framework for Understanding*, Addison Wesley Publishing Company(1988).

44. Pan, J. Y. C., and J.M. Tenenbaum. "An intelligent agent framework for enterprise integration." *IEEE Transactions on Systems, Man and Cybernetics, Special Issue on Distributed Artificial Intelligence*, Vol. 21, No. 6 (December, 1991), pp. 1391-1408.

45. Parunak, H. V. D. *Manufacturing experience with the contract-net*. Huhns (1987), pp. 285-310.

46. Rumbaugh J., Blaha M., Premerlani W., Eddy F., Lorensen W, *Object-Oriented modeling and Design*, Prentice-Hall (1991).
47. Searle, J. R. *Speech Acts.* Cambridge University Press (1969).
48. Searle, J. R., and D. Vanderveken. *Foundations of Illocutionary Logic.* Cambridge University Press (1985).
49. Smith, R.G., and R. Davis. "Frameworks for cooperation in distributed problem solving." *IEEE Trans. Syst. Man Cybern.*, vol. SMC-11, (1981), pp. 61-70.
50. Steiner, D.D., and D. E. Mahling. "Human computer cooperative work." In *The Proceedings of the Tenth Workshop on Distributed Artificial Intelligence,* Bandera, TX, (October, 1990).
51. Sycara, K. R. "Argumentation : Planning other Agents' Plans." In *Proceedings of the 11th International Joint Conference on Artificial Intelligence,* Detroit (1989).
52. Van Assche, F., B. Moulin, and C. Rolland. *Object-Oriented Approach in Information Systems.* Proceedings of the IFIP TC8/WG8.1 Working Conference, Elsevier Science Pub (1991).
53. Von Martial, F. *Coordinating Plans of Autonomous Agents.* Lecture Notes in Artificial Intelligence, Springer Verlag (1992).
54. Wilensky, R. *Planning and Understanding.* Addison Wesley (1983).
55. Wilkins, D. *Practical Planning.* San Mateo, CA., Morgan Kaufmann (1988).
56. Winograd, T., and F. Flores. *Understanding Computers and Cognition: A New foundation for Design.* Norwood, NJ., Ablex (1986).
57. Zachary W. W., and S. P. Robertson, Eds. *Cognition, Computation and Cooperation.* Ablex (1990).

AUTHOR INDEX

A
Aliev, R.A., 99
Aminzadeh, F., 29

B
Barak, D., 45

C
Chew, G., 109
Cloutier, L., 261

D
Ding, H., 165

G
Ghenniwa, H., 229
Gupta, M., 121, 165

H
Hall, L.O., 1
Hatono, I., 217

J
Jamshidi, M., 45
Jin, L., 121

K
Kamel, M., 229
Kandel, A., 1, 109
Kelsey, R., 45
Kristjansson, E., 45
Kumbla, K., 45

L
Lang, Z., 201
Langholz, G., 109
Li, Y., 153

M
Marchbanks, R., 45
Moulin, B., 261

N
Nikiforuk, P.N., 121

S
Scarberry, R. E., 201
Schneider, M., 109
Simaan, M., 201

T
Tamura, H., 217

Z
Zhang, X., 153
Zhang, Z., 201

SUBJECT INDEX

A

Aptronix-FIDE, 48
antecedents and consequents, 46
artificial intelligence, 218
    distributed, 230
artificial neural networks, 153
automated guided vehicles, 217
autonomous agent, 245

B

Bell Helicopter's FULDEK, 48
behavior diagrams, 277

C

case-based reasoning tools, 1
certainty factor vector upgrading, 209
coarseness measures, 208
coherency knowledge,
    incompleteness of, 234
cold-junction compensation, 81
computational neural architectures
    for control applications, 121
computer-supported cooperative
    work, 263
conclusion clause, 114
concurrency, 250
confluence operation, 187
contract-net protocol, 264
coordination control, 234
coordination structure, 234
cross-layer connections, 121

D

DAI environment, 245
Dempster Shafer's "mixed
    initiative" approach, 30
Difference Summation Radial
    (DSR), 174
decentralized structure, 234
direct inverse control, 130
disposition, 11
distributed artificial intelligence, 230
dynamic backpropagation, 146

E

EMYCIN, 6
ESPLAN, 99
EXPERT, 6
entity life-cycle, 277
executable clauses, 113

F

Flexible Manufacturing Systems
    (FMS), 217
FMS simulator, 222
fuzzification/defuzzification, 46
Fuzzy Associative Memories
    (FAMs), 51
Fuzzy Operation Form (FOF), 178
    acceleration control, 75
    activation, thresholding
    and nonlinear, 183
    aggregation, 183
    control, 46
    delta rule, 187
    expert systems, 30, 99
    uncertainty in, 8
    IF-THEN rules, 1
    inference network, 37
    logic, 30, 46
    mapping, 178
    neural networks, 165
    models, 187
    inclusions and limits, 187
fuzzy vector projection, 178
fuzzy-sets, neural networks, 187

G

GUITAR, 288
genetic algorithms, 1
global segmentation database, 207

H

Helikon's IDL-747 imbedded fuzzy
   controller board, 48, 49
Human-Computer Cooperative
   Work (HCCW), 270
hierarchical architecture
   intelligent control, 16
   timed transitions, 222
   multilevel, 234
hybrid concept, 13
hybrid intelligent systems, 14

I

Inference Engine, 6
ISDEL Algorithm, 205
image segmentation, 202
information processor, 167
initiate-meeting scenario, 277
intelligent agent, 245
intelligent control hierarchical
   architecture, 16
interdependency knowledge,
   incompleteness of, 234
interdependency, 250
interleaving Ps-Sr-Cr, 245
intra-layer, inter-layer and
   recurrent connections, 121

J

JIT, 221

K

knowledge and numeric
   information, 209

L

laser beam system, 48
lateral connections, 121
learned experiences, 167

M

Michigan Intelligent Coordination
   Experiment, 230
MYCIN, 6
matching factor, 114
membership function, 46
meta-knowledge, 6
model train, 75
multi-level hierarchical or
   market, 234
multiagent planning, 269
multiagent system (MAS), 262
multilayered feed forward
   neural networks, 123

N

mutual relationalship, 167
NeuraLogix FMC board, 48
neural networks
   artificial neural networks, 153
   fuzzy neural networks, 165
   multilayered feed forward
      neural networks, 123
   recurrent neural networks, 142
   tapped delay neural
      networks, 130
neuro-fuzzy logic, 178
neurons, 167
neurons, 123
non-chlorofluorocarbon (CFC) air
   conditioning and refrigeration
   system, 79
nonlinear activation operation, 167
plants, 146
novelty filter, 155

O

object modelling, 287
object structure, 288
object/agent interaction, 291
oil exploration, 33

oil industry, 30
on-line scheduling, 219
oil refinery plant, 99

P

Petri nets, 217
PROSPECTOR, 30
Product Summation Monotonicity (PSM), 170
power generating system, 56
power system generation, 56
premise clause, 114
preprocessing, 109
principle of graceful degradation, 209
propagation networks, 123
proportional-integral controller design, 84

R

R-List, creating, 114
Rhino robot, 63
reachability, 250
real-time architecture, 49
recurrent neural networks, 142
relative degree, 142
robotic manipulator, 63
rule base, 221

S

SEEK, 6
SPERILL-II, 9
System Intelligence Quotient (SIQ), 1
seismic data, 153
smart manufacturing products, 79
society, 250
soft-logic for FNNs, 165
somatic operation, 167
spatial data structure, 203
supervised leaning, 157
symmetric ramp function, 138
synaptic operation, 167

T

THEISESIAS, 6
Theory of Coordination, 231
Thermoelectric Devices (TEDs), 79
Togai's SBFC, 48
tapped delay neural networks, 130
texture energy measures, 212
thermoelectric device-based refrigeration system, 79
trace editing, supervised, 157
 unsupervised, 158
train. fuzzy control, 75
transition diagram, 277
two or three dimensional image data, 202

U

uniformity measures, 208

Please send me information on:

- ☐ *Fuzzy Logic and its Software Applications*
- ☐ *Fuzzy Logic and its Hardware Applications*
- ☐ *Fuzzy Logic Control and its Applications*

Name: _____

Institution: _____

Address: _____

_____

Phone: _____ **FAX:** _____

E-Mail: _____

CAD Laboratory
P.O. Box 14155
Albuquerque, NM 87191-4155 USA
Phone: (505) 277-5538 FAX: (505) 291-0013

Please send me information on:

- ☐ *Fuzzy Logic and its Software Applications*
- ☐ *Fuzzy Logic and its Hardware Applications*
- ☐ *Fuzzy Logic Control and its Applications*

Name: _____

Institution: _____

Address: _____

_____

Phone: _____ **FAX:** _____

E-Mail: _____

CAD Laboratory
P.O. Box 14155
Albuquerque, NM 87191-4155 USA
Phone: (505) 277-5538 FAX: (505) 291-0013

Place Stamp Here

CAD Laboratory
P.O. Box 14155
Albuquerque, NM 87191-4155
USA

Place Stamp Here

CAD Laboratory
P.O. Box 14155
Albuquerque, NM 87191-4155
USA